I hope you enjoy reading this
autobiography with all the historical
events that occurred before and
during my life times.

Joe

WHEN I GET BIG,

I'M LEAVIN'

The 86-year journey of an immigrant's son
from an Appalachian coal mining town
to ultimate success

An Autobiography

By

Joseph T. Diviak

ISBN 0-7414-1041-9

Published by:

INFINITY
PUBLISHING.COM

519 West Lancaster Avenue
Haverford, PA 19041-1413
Info@buybooksontheweb.com
www.buybooksontheweb.com
Toll-free (877) BUY BOOK
Local Phone (610) 520-2500
Fax (610) 519-0261

Printed in the United States of America

Printed on Recycled Paper

Published July, 2002

Preface

The title of this book came from the phrase I used the first time I rushed across the railroad tracks in front of our house in Caretta, West Virginia and saw the bodies of two crushed coalminers. They had been brought out of the mine in a small underground coal car, covered by a tattered piece of canvas, waiting to be declared dead by the company doctor.

As children, we saw a lot of men brought out of the mines after accidents. But this one was a terrible picture. What made it even worse were the cries of the wives and children. I was eight years old and, as I stood by that coal car, I kept saying, "When I get big, I'm leavin'!" And I did.

We saw firsthand what the coalmines did to our fathers, relatives, and friends. Most of the men were all stooped over and were considered very old at only 45 to 50 years old. We saw how the coal companies took advantage of the workers—treating them like slaves in the mines, taking advantage in the company store. We had lots of reasons for "leavin'" when we became adults.

Dedication

This book is dedicated to my parents, Joseph Emory Diviak and Helen Kalany Diviak, who escaped to America from Austria, Hungary in October 1906. They left behind years of discrimination and apprehension that had been endured by their Slovakian ancestors.

To them I owe everything for their courage to come to a strange land in order to rear a family in a free country.

They wanted the best opportunities for success for their children yet to be born and enjoy life to the fullest.

All of this was done for us nine children, and to them, using an old mountaineer's phrase, "we are very beholden!"

Credits

To my first wife of 45 years, LaRue Cummins Diviak, who encouraged me in every possible way. She was behind me a hundred percent in all the changes and decisions I had to make in order to have a very successful career in electronics and other fields of endeavor. This helped to write this book and the history of our family.

Credit goes to my second wife, Ruthie Horton Diviak, for laying the groundwork when she wrote her book, A Courageous Woman. All I had to do was follow the same procedure. She encouraged me to complete my book, begun many years before.

Contents

CHAPTER ONE

Escape from Austria, Hungary

It was getting dark in Vienna, Austria on that September evening in 1906. A visitor and his companion huddled in a recessed doorway, trying to avoid the cold wind and stay warm.

Added to their concern was the shadowy figure down the street. It was clear they were being followed. When they stopped, so did the figure, pretending to look in the show window of the shops that lined the street.

They were supposed to meet someone secretly, someone who was supposed to be bringing them a pair of fake passports, someone they did not know. A friend in Bratislava had made the arrangements. Each night for the past two nights, the couple had been at the appointed place at the appointed time, but no one met them there.

Finally, on the third day, as they were having lunch at a café, a well-dressed man carrying a folded newspaper in his left hand, sat down across the table. "It is wonderful to see you again, Helen and Joseph!" The man said, smiling at the couple. "I have a letter for you from your uncle and he is well." He handed Joseph an envelope then the stranger ordered a cup of coffee while Joseph quickly read the letter and slipped the passports into his pocket.

"Oh, yes," Joseph said. "I have a letter for our uncle, if you will be so kind to deliver it to him?" He handed the stranger another envelope. In it was the koronas for payment for the passports.

The stranger, still smiling, put the letter in his coat pocket and whispered, "I had to follow you for two evenings to make certain you were not being followed. You will not have any problems if you do not attract attention. Go with God and may He be with you." The stranger rose, shook hands with the couple, and left.

Helen and Joseph breathed a deep sigh of relief. They finished their lunch then returned by a different route to Helen's brother, Eugene Kalany's home to pack and catch the next train to Bern, Switzerland. Their journey to America had begun.

Eugene and his wife, Anna, had two children, Eugene, Jr., who was four years old, and Anna, who was two. Twice a year, the children were taken to visit their grandparents in St. Peter by train, which the children thoroughly enjoyed.

One day, Eugene and his family were visiting and, as planned, left the children with their grandparents. A few days later, Joseph and Helen, each carrying one bag and a child, were at the train station in Bratislava. The children were very cute and all smiles, especially at the uniformed Hungarian policemen at the station.

As Helen and Joseph boarded the train to Vienna, each lifted a child in their arms. The police were so taken with the smiling, friendly children that they forgot to check Helen and Joseph's regular passports. Helen and Joseph took the children back to their parents in Vienna and began their search for the man with their new passports. Their plan was to travel from Bern, Switzerland to Paris, France then to Bordeaux, France, where they would board a ship bound for America. The ship was to be in port, ready to depart when all the passengers had arrived.

Joseph leaned back in his seat and reflected on what a life it had been for the last twenty years. His father, John, was a flax farmer. The farm was a fifteen-minute walk from their home in the village. All the houses, with their gardens, were clustered around the church, for mutual protection from bandits and thieves.

But the best benefit by far was the fellowship among the people of the village. It was a most enjoyable life, especially at baptisms, weddings, and holidays. All were good reasons to have parties. They helped each other through illnesses, their harvests, and hard times

But overlooking this idyllic village were the gendarmes—the Hungarian police—stationed in each village to "control" the people as they saw fit. They would often beat up boys who were caught outside after 9 P.M. They were the law and there were no appeals.

Joseph's father, John, had served in the Austro-Hungarian Army during a war against Italy in the 1860's. He had been a commander of a "ball and chain" canon company that guarded the bridges across the Danube River. Two canons were placed side-by-side, about ten or twelve feet apart, pointing across the bridge roadway. Two cannonballs were linked together by a chain, one ball in each canon.

The cannons were aimed about five feet above the roadway. When both canons were fired simultaneously, the chain stretched between the balls and cut everything in their path in two: men, horses, wagons, or whatever. There were several canons for each bridge or roadway. In one battle, the Danube actually ran red when thousands of Italian soldiers tried to cross the bridges.

John's best friend in this battle was Russian officer mercenary, Ivan Petrov. The two spent a lot of time together planning battles and just talking. After the war was over, John and Ivan exchanged addresses and corresponded for many years. Ivan told John that he wanted to be notified if John had a boy child after he and his childhood sweetheart married. When Joseph was born, after his four sisters, John wrote to tell Ivan in Russia.

A few weeks later, a crate arrived at John's home. It contained a Siberian wolf puppy! The puppy was named "Sheppie" and he and Joseph became inseparable pals. The villagers marveled at how Sheppie behaved and always protected the little boy.

The gendarmes usually traveled in pairs, but even so, they stayed clear of Shep and Joseph. They seemed to be a bit uneasy around the dog. But they soon grew to like Shep.

When the Gypsy Caravans would travel through the countryside and set up their tents and carnivals outside the village, Shep would spend his days and nights inside the large barn on the farm and Joseph and John would spend a lot of time with him. The reason for this was that the Gypsies traveled around, making note of all the things that could be easily taken. They always left very early in the morning and everything they could take went with them, including chickens, pigs, etc.

In later years, Joseph told his own children many stories of the twenty years he spent in Slovakia.

When he was twelve, he began an apprenticeship as a linen weaver. He served four years as an apprentice, then, at age 16, he had his own loom at home. His mother, Maria Skapik, was a wonderful mother and made sure that her daughters, Pauline, Anna, Agnes, and Julia, learned everything possible about homemaking and gardening.

Joseph purchased thread at the flax mill to make linen. He wove the linen in large rolls and traded for money, boots, clothes, or whatever he could, for the family. The barter system was used a lot in the villages in those days.

When Joseph left St. Mikulos, he left behind more than 30 each pairs of linen slacks, shirts of different colors, many pairs of boots and hats he had bartered for over the four years.

When he was twelve, he was recruited into a band formed by the local bandmaster. Of the thirty who joined originally, only 13 were kept. Joseph was one of them. He played lead coronet in the band, at church functions, dances, wedding, and such. He was always paid in "gold pieces" –

koronas – which he hid and saved and later used to pay expenses to go to America.

When Joseph was 16 or 17, he met Helen Kalany at one of the joint church activities held by the churches in St. Peter and St. Mikulos. It must have been love at first sight because the two met as often as they could, under the watchful eyes of their elders.

Helen Mary Kalany was the daughter of Florian Steven Kalany and Barbara Centilli. Florian was a supervisor in the Forest Rangers. He and his family lived in a nice, large home given to him by the government. Like John Diviak, he had a second skill—he built beautiful pieces of furniture.

Helen and Joseph completed all the schooling available to them in their area. Helen served as a messenger between forestry stations until she escaped to America.

When Joseph turned eighteen, a government official told him that, when he turned twenty, he would have to serve two years in the Austro-Hungarian Army, and that he would not be permitted to marry until after his two years was up. He was also told that it would not be wise for him to marry Helen, because she was the daughter of a government official and he the son of a farmer. Helen and Joseph were shocked—the old caste system!

Some older friends who had already served their time in the army had told Joseph that the Austro-Hungarian officers and non-commissioned officers had mistreated them the entire time. This perhaps planted the idea that America was their only hope of being together. Before long, Helen and Joseph quietly began planning their escape. They had heard much about America from friends who had immigrated there.

The Diviaks and Kalanys belonged to the Reformation Church, as did most of the inhabitants of their small village. This church, along with the Hussites, was formed before and after the burning at the stake of a Catholic priest by the

Roman Catholic Church for heresy in 1434. John Huss, the priest in question, was a Bohemian reformer whose worst 'sin" was translating the Bible to the Slovakian language so the common people could read it.

However, at that time, only a priest was permitted to read the Bible and could then give the people the "approved" interpretation, according to Church law. One Cardinal and two Bishops were sent from Rome to conduct the trial of the hapless priest, after various investigators from Rome made their reports. They were also present at the trial. John Huss was given a chance to recant, but he would not, so he was burned at the stake.

Many were the stories that Helen and Joseph later told about their lives in the villages in Slovakia. One in particular that stands out is about the Easter Parade of young maidens who would lead the congregation up a flight of eighteen steps and into the Church. According to the legends, a girl who was not a virgin would not make it to the top because God was watching. Tales from previous years told of a few who made it to the tenth or eleventh step and fainted.

These tales put considerable pressure on these young girls. But, as Helen and Joseph told it, none of the girls walking up the steps ever had a problem during the time they were there. Of course, they pointed out, it was always nice to have distant relatives that had to be visited that particular Sunday. There were also more colds and fevers during that week as well!

A few of the mothers started a petition to have young men of the same age as the maidens follow the girls up the steps. All had to be virgins and all had to be accounted for in the area. The following Easter, there was a blinding snowstorm during Easter, so the parade was cancelled. The year after that, it was mercifully forgotten, much to the relief of not only the boys, but also the parents of girls who had no relatives outside the area!

As Joseph and Helen leaned back in their seats on the train, they felt fear, yet a bit of hope. They arrived in Bern and their passports were checked at the railroad station. There was also a telegram awaiting them with the police there, urging them to return to Vienna. Joseph, with a couple of coins in his palm, shook hands with the Swiss policeman and asked, "Do we have to go back?"

The policeman smiled. "I am telling you to go." He winked at Joseph. Helen and Joseph caught the next train to Paris.

That, too, proved to be a long and tiring trip, trying to sleep sitting up on the train, but being young and in love, they managed. When they arrived in Paris, a gendarme looked at their passports and shook his head from side to side. "We have a telegram for you to return to Vienna from the police there." Joseph, again with gold coins in his hand, shook hands with the gendarme, who said, "Do as you please but don't stay here!"

The couple boarded the next train to Bordeaux. After still another long, tiring train ride, they finally arrived in Bordeaux. They had no problems there as they boarded the ship, a two "smoke stacker" ship, considered small even in those days. When they boarded, they found more than 200 other people aboard who were also "escaping to America".

At sundown, a group of gendarmes accompanied by a ratty-looking individual came aboard. The man kept yelling and pointing to some of the people standing on the upper deck of the ship. It turned out that the man was a bounty hunter and was demanding that all of the 'escapees" be arrested.

The captain of the ship prevailed, telling the police official in charge that there was too much confusion aboard the ship and that some people had already gone to bed. He asked the official if he would "please return at eight the next morning".

The ship pulled out at 3:00 A.M. and was far out in the ocean by 8:00 A.M. Later, the captain received a message that the bounty hunters body had been found floating in the water by the dock and asking that, if anyone knew anything about it, the police would appreciate any information they might give. Of course, no one aboard the ship knew anything about it

The men and bigger boys were put in the large wardrooms in the stern of the ship, while the women, girls, younger boys, and babies were put in the forward wardrooms. Freight was in the center section. It was a very slow freighter and would take two weeks to reach New York City. On the way across the Atlantic, they encountered severe rain and gale force winds most of the way.

About two-thirds of the way across the Atlantic, many people had gone on deck during a lull in the rain. Patches of fog were rolling in. Suddenly people began to scream and the ship's horn began to blow frantically. A large steamship was coming from their left, across their path and was by then very close to them. The rough seas and high wind and swells tossed both ships around like they were out of control.

The freighter's crew was readying lifeboats. There was no time to get the people still below deck to the top deck. The oncoming steamer began to turn to its left while Joseph and Helen's ship drove full steam ahead. At the last moment, the captain turned the freighter sharply to its left to get the stern out of harms way.

At the very instant that everyone on deck was sure the two ships would collide, a huge wave and its following swell came between the ships and the bow of the oncoming steamer was turned away. They passed without touching. It must have been quite a scene: both ships' horns blowing, people screaming, and the wind howling. Later, Joseph said he was so surprised when the two ships didn't collide. Later they found out that the other ship was a coastal freighter that had come across their course.

Joseph came through the voyage without any sign of seasickness at all. In fact, he ate all the pies and cakes he wanted. Helen, on the other hand, had motion sickness and, along with most of the other women aboard, was ill most of the time. Helen and Joseph met on deck several times, but never for long because the weather was so bad.

One of the prettiest and highly emotional times was when everyone on deck finally saw the Statue of Liberty. The deck erupted with cheers as tears of joy streamed down their faces. The joy these immigrants felt was so deep, it was hard to describe. At last, Joseph and Helen were free. Their well-planned escape had become reality.

There were no problems going through Customs at Ellis Island since there was no quota limiting the number of immigrants entering the United States in 1906. Joseph was careful to make certain their names were spelled correctly. He had heard aboard ship that due to the number of people entering the United States and the time allotted to the processing of these people, often names were unintentionally changed, partly due to the time constraints and partly due to the language barrier.

Helen and Joseph boarded a train in New York City, bound for Cleveland, Ohio, where they had friends who had immigrated a few years earlier.

Helen and Joseph were married by a magistrate and rented a house. Joseph had found a job in a saloon in Cleveland but only worked there for a few months. He spoke several languages and that helped him in his job.

CHAPTER TWO

Starting a Family

Helen was expecting a baby. Joseph immediately began to look for a better job, one that paid more money and would allow him to be home every evening with his family. A friend at the saloon told him that good jobs were available in Wheeling, West Virginia, in the steel mills.

Joseph took the train to Wheeling where he easily secured a job. He found a house to rent, then returned to Cleveland and brought his pregnant wife back to Wheeling. Theresa was born on October 11, 1907.

He worked on the loading dock; every day, he loaded hundreds of wooden boxes of large bolts and nuts on the wagons backed in to the dock. It was hard work and a rough ordeal for Joseph, but he was a large, strong man and he did the job easily.

Joseph had worked at his job for two years. He had a happy family and things seemed to get even better once they had made a few friends. Mary was born on January 31, 1909.

One of the drivers who handled the horse drawn wagons backed in to the dock every day was a foul-mouthed man. He would back his team and wagon up, then say to Joseph, "Load it up, you son-of-a-bitch!"

The others would laugh, but Joseph didn't understand much English, so he would smile and nod to the man, then load the wagon to capacity quickly. But it bothered Joseph that he didn't know what the man was saying to him.

The next morning, Joseph asked a German friend, "Vas is los mit der 'son-of-a-bitch'?" He went on to explain that one of the drivers called him that and the others laughed.

The friend told Joseph it meant "swinehund", which is the equivalent of 'son-of-a-bitch'. Joseph was quiet.

The next morning, the foul-mouthed driver backed in his wagon, came up on the dock and said, "Load it up, you son-of-a-bitch!" to Joseph. Without a word, Joseph grabbed the man by the seat of his pants and the back of his neck and rammed his head into a tall stack of boxes. Some of the bolts and nuts fell on the man. Leaving the man lay there, Joseph turned and walked off the dock. On his way home, he stopped by the makeshift hospital and told the doctor that he would pay the man's bill. Then he went home.

That evening the foreman came to Joseph's house. He told Joseph that the other workers had told him how Joseph had taken a lot of abuse from that driver and that they felt what Joseph did was justified. He asked Joseph to come back to work. But Joseph told the foreman that he could not work for a company that would hire people like that. He and his family were leaving Wheeling.

He had heard that the coal company in Berwind, West Virginia was hiring coal loaders. Again, Joseph boarded a train, secured the job, and moved his family. He had found a larger than usual house and had met two men, Frank and John Marfi, twins from Hungary, who needed a place to live, so Joseph rented the house and the twins moved in as boarders.

It wasn't long until Frank and John became "uncles" to the children. They had come to America to make their fortunes. They planned to move back to Hungary, buy farms, and marry their sweethearts. To this end, they sent money home every month. They were a great help around the house and helped with the children. Anna was born on January 6, 1911 and suddenly they had a house full!

Two years later, the Berwind Coal Company ran out of orders for coal. Joseph, Frank, and John went job-hunting together. Their search led them to the small town of Caretta, about twelve miles from Berwind, still in McDowell County. There they found jobs and a house large enough for all of them.

The town of Caretta was named after the owner of the mine, George Carter, and his wife, Etta. It was a small town but growing. There was a 48-inch seam of coal at valley level under the mountain and an untouched 6-foot seam below that.

The first house the Diviaks lived in was a stucco frame house, number 48, where Helen was born on January 1, 1913. A year later, the family moved two houses down the hollow, number 46, which was a better arranged house and Joseph, Jr. was born there on May 5, 1915. Joseph, Jr. was baptized on May 15, 1917, in the Catholic Church in Berwind, because there were no Reformation Churches in the area.

During those years, there were a lot of activities in the saloons. Beer could be bought in containers brought in by customers and raffles were popular.

When little Joe was eighteen months old, his mother held him up so that he could reach into a hat to select the winning raffle ticket. He drew number thirty-five, which later turned out to be the winning ticket for an 18-carat gold, 18-jewel, double cased pocket watch. Later, his sister Mary won a shotgun and a cow, both of which were sold. Joseph Jr. still has his watch.

Joseph Senior was so proud of his children and having a son was the crowning glory—a son to pass on the family's heritage.

In the back yard of the house, nestled against the hillside, was a large concrete dome-shaped oven, with a firebox underneath and the oven above. All the neighbors shared this oven, a kind of community affair. Helen would bake large loaves of bread there every Wednesday and pastries on Saturday.

(May 25, 1915)
My mother, Helen Diviak, holding me in her arms at age 3 weeks old, after christening me Yosef Diviak, 25 May, 1915. My father, Joseph, and my godmother, Erma Gombas, are with her. Caretta, West Virginia

Mary (Skapek) and John Diviak (1920) in Slovakia

(May, 1915)
My father, Joseph E. Diviak (center) and our boarders, Hungarian twins Frank and John Marfi, "hamming it up" on a Sunday, — with tools, carbide lights on caps, and a miner's lunch pail.

The coal companies would compete for orders from northern cities. At times, the mines would only work one or two days a week. Eventually the Diviak family moved to Eclipse, West Virginia (no longer on the map). Julia was born there on September 6, 1917. The clouds of war (WWI) hung over Europe and America, affecting the economy of our country.

The draft dug deeply into the younger coal miners and the older miners were sent back to work the mines. Years later, Joseph was asked why he had not been drafted.

"During checkup, I no yell loud enough," he replied.

We all laughed, thinking he was kidding, but in the absence of telephones, yelling loud and far was their primary way of communicating. However, the main reason may have been that Joseph held the best coal loader's record at the mine.

When Joseph, Jr. was three years old, in 1918, the influenza epidemic was taking a deadly toll in Europe and America. He caught the flu, along with his mother and the other children. Joseph, Sr., Frank, and John never got it.

One night, Joseph had a very high fever. Every breath was a struggle and he began to turn blue. His father picked him up, turned him over, and gently shook the little boy. Joseph, Jr. expelled a large amount of phlegm and immediately began to breathe normally, his color a healthy pink again.

The care of the family fell to Joseph, Sr. Fortunately he knew how to cook good nourishing meals, which he did, and still worked in the mines until all the children and his wife were fully recovered.

A move to Pineville, West Virginia was cut short and instead the Diviaks moved to Taplan, near Logan, on the Guyandotte River. Their house sat on the banks of the river and young Joseph loved to play under a large sycamore tree there, watching the fish swim by on a sunny day.

A "chung-chung-chung" sound was strange to Joseph. (Many years later, Joseph, Jr. heard that same sound again, at a sugar cane plantation and discovered it came from a single-piston Fairbanks-Morse engine with a belt drive attached to a cane crusher.) It was also used to grind corn and wheat. Sears, Roebuck sold hundreds of them and many found their way into boats.

The next morning, Joseph, Jr. went out to see the boat and the people on it. The people introduced themselves: Sam Hunter, Sr., his wife, Agnes, and their 5-year-old son, Sam, Jr. At last, little Joseph had a playmate other than his sisters!

Sam, Sr. told how he and his family came up the Ohio River, then the Kanawha and Guyandotte Rivers every May. They stayed for four months. Sam did a lot of carpentry work for the coal company, repairing the homes and building all summer. In the fall, they would return to a community on the Mississippi River to spend the winter in colonies with other 'boat people'. It turned out that the sycamore tree was their favorite place to tie up for the summer.

John had been born on February 24, 1919. That made five girls and two boys. We were fortunate to be able to find a house large enough to handle nine people, counting Frank and John, our boarders.

Sam, Sr. spent his days working for the coal company while his wife, Agnes, took the little skiff tied to the back of the houseboat and paddled to an island near the cliffs on the other side of the river. There she planted corn, beans, tomatoes, onions, and so forth. As her harvest ripened, she was busy canning the vegetables.

The houseboat had two bunks on each side of its one large room. It was about nine feet wide and twenty-eight feet long. The walls were covered with shelves from floor to ceiling, except for one window on each side and a door at each end. There were screens on the windows and doors to keep out mosquitoes and flies.

In one corner of the room, there was a small step-in closet, which was the bathroom. In the very center of the main room was a coal-burning stove, for heat and cooking, and a table with a long bench on each side. Near the back door was a smaller table that held the water buckets. Utensils hung on a small portion of the wall.

One evening, Sam and his family went into town to spend the night at a friend's home. Sam had made one huge mistake—the bow of the boat was tied low to a root of the sycamore tree. During the night, a heavy rainstorm upriver caused the river to rise. In the morning, the front end of the boat was under water, while the stern was up in the air.

Joseph, Sr., Frank, and John ran down to the boat. One of them dived into the river and untied the bowline. The boat righted itself, full of water but still afloat. The men retied the bowline, then they started bailing the water out of the boat. Meanwhile, Sam and his family returned, grateful that their boat had been saved. The boat was scrubbed clean and no damage was done to the engine.

During the bailing process, the men found a number of large perch and bass trapped in the boat, and these, along with close to a bucket of crawfish, were salvaged. That night, there was a cookout on the riverbank. Neighbors brought covered dishes of food to go with the spicy cooked crawfish and fried fish.

That fall, as the Hunters were preparing to leave for the Ohio River, they came to our house for supper. Joseph, Sr. asked Sam if the cove where all the boat people tied up was safe and whether the other people were nice.

"Mr. Joe," Sam answered, "We are safe because we all look out for each other on the river. If we happen to get a family that's bad or troublemakers, they will wake one morning to find themselves twenty or more miles downriver, because someone cut his anchor line during the night."

The next morning, young Joseph ran out on the back porch and the houseboat was gone. It had silently drifted away during the early morning hours, headed for the Ohio River. Joseph, Sr. said he had heard the "chung-chung-chung" of the boat's engine faintly as it drifted clear of the town early that morning.

An elderly man, bent with age, with a long beard and long white hair, would walk past our house with a burlap sack over his shoulder. The older kids told us that he carried children off into the woods in his sack if they were mean. That frightened Joseph, Jr. and a few of the other kids.

After the early fall rains, large rafts of logs were floated down the river to the sawmill. Each raft was huge and long. There were two men on the front manning a twenty-foot paddle and another man at the rear of the raft steering to keep the raft in the middle of the river. A large tent and bunks were in the center of the raft. They tied up with a steel cable at night. All the children would gather on the banks to watch the rafts go by.

One day, the old man with the sack was walking past and saw the large dray backed up to our porch being loaded. We were moving. The old man called out to Joseph, Jr., saying, "I won't hurt you. I am your friend! I wish you long life." He handed little Joseph a bag of peppermint candy and said, "Good-bye."

Joseph, Jr. thanked the man, shook his hand and said good-bye to him.

The move to Blair lasted only a few months, then the Diviaks moved to Jeffery, in Boone County. Whenever the family moved from one town to another, Joseph, Sr. always rented a boxcar from the North Fork & Western Railroad. He and Frank and John took all of the household items, and the family milk cow, with plenty of hay into the boxcar. Joseph usually stayed in the boxcar as well. He was always well supplied with food, coffee, and water. It was his way of insuring none of their property was "lost". He carried a shotgun and a Smith & Wesson .38 revolver, so he felt safe.

The rest of the family rode in the passenger cars, slumped in their seats, with no air conditioning. The windows were wide open and the coaches filled with smoke and dust as the train passed through tunnels.

The toilet facilities on the train consisted of a washbasin with water faucets, a mirror, and a toilet. If you lifted the cover of the toilet and pressed the flush lever, the trap in the bottom of the commode opened and you could see the railroad ties zipping by in a blur. Not very sanitary, but practical!

In the end, the move to Jeffrey was successful and everyone arrived safely.

CHAPTER THREE

Strikes and Problems

Joseph Junior tells the story from here on as he remembers it.

I was so impressed at the size of the house we were moving into. It had once been a farmhouse. The coal company had purchased along with the surface and mineral rights. The coal company built a number of smaller houses on these former farms to house their workers.

Our house stood on a corner facing Main Street, which curved around the house along the Little Coal River. The streets in those days were unpaved. The house also had a large wraparound porch that went from the front of the house, down the side that faced the river and around the back.

It was a beautiful setting, with thirteen apple trees around it. These yielded several types of apples, which all of us enjoyed. A four-foot high wire ornamental fence enclosed the house, trees, the washhouse, and the chicken house. It was certainly a well-kept place.

There was a large bunkroom on the second floor of the house with lots of closets and plenty of screened windows, which was certainly nice for our large family.

The house sat on four-foot high piers, which was a nice place for us children to play. Out the back door, on the porch, was a well with a three-foot high box around it and a cover. A pulley hung from the ceiling of the porch and held a rope with a bucket attached. The outhouse was near, in the bathhouse. Water was heated on the coal-burning stove for bathing. There were no bathhouses at the mines.

Soon after we moved in, four coal miners from Bulgaria asked to board with us. They had been living in a small shack at the edge of town. One we called "Cook" did all the cooking for them and packed their lunches. He was getting tired of working in the mines and doing all the cooking too.

Since we had ample room in the upstairs bunkroom and they all promised to help with the cooking and household chores, they moved in with us. From the outset, we all enjoyed them. They were young and full of pranks and enjoyed games, put on plays and were a lot of fun.

Their names were John Lisconoff, Andy Zerinko, Anthony Tomov, and "Cook" George Petroff. They loved our family and helped us in every way they could. They became great friends with out twin boarders, Frank and John Marfi. The four of them also had come to America to make their fortunes, then return home, buy farms, and marry their childhood sweethearts—the same goal the Marfi brothers had.

To help provide food, they fished and bought lots of meat. They loved rabbits and one day, "Cook" brought home three rabbits, a buck, and a two doe—Belgian Browns and the largest rabbits we had ever seen. Some of the men went to the sawmill and carried home bundles of slats to make a picket fence around the storage room behind the washhouse.

The storage room was the size of a two-car garage – about 20 feet by 20 feet – and they cut a little arched doorway into the wall. Then they built lots of boxes that resembled small doghouses for the rabbits. Within a few months, we were enjoying stewed rabbit or fried rabbit with brown gravy and mashed potatoes, canned beets and such at least once a week. It seemed the rabbits took the admonition to "go forth and multiply" very seriously.

Feeding the rabbits was not a problem. Along with sweet clover, there was "chop" bran made of hulls from wheat in hundred pound sacks, which we made into a mush. That was one of their favorite staples.

When the rabbits were slaughtered, great care was taken to skin them in such a way so as not to damage the hides. The hide was turned inside out and dried on a thin wooden form in the barn. Later, all the hides were packed and sent to Sears with one of their agents. The money went into getting "rabbit food".

We also had three large hogs by fall every year, forty or more chickens, four or five mallard ducks, and two or three geese. Turkeys were unheard of in those days. The food we ate was similar to what my parents grew up on in Europe.

One of my chores was to take my little wooden wagon and pick up all the apples that had fallen from the trees during the night or after a rain. Apple pies, stewed apples, and apple fritters were wonderful! The hogs and chickens got their share, too. The hogs especially loved the overripe apples.

Our ducks did not like to hatch their own eggs, so Mother put them under a "setting hen". It must have been a terrible trauma for that hen when the ducklings headed for the river next to our house. She would screech and flutter over the water, trying to get them to come back on dry land. After a while, the hen would resort to scratching around the river bank, pretending to find food, while the ducklings had their daily fun time "ducking" into the water and "peeping" with glee.

There was another farmhouse behind ours where a farmer, Worley Chambers, and his ailing wife lived. Worley also worked in the mines. His son, Worley Junior, was my age (six) and we often played together. Our fathers cut some long thin willow saplings and made fishing poles for us. The fishing line was black sewing thread and the hooks were made by bending the head of a straight pin to make a loop for tying the line, with a curved hook at the other end. A cork from a medicine bottle made an excellent float.

We dug our worms out behind the barn and put them in an old tin can. We'd gather our poles and a bucket and head for the "Blue Hole" near our homes on the Little Coal River.

We caught a lot of perch, trout, and bass, but lost a lot of them since there was no barb on our hooks. When we saw the cork dip under the water, we learned to pull quickly and throw the fish up on the bank behind us. Then we would fall on the fish, grab him and put it in our bucket that was half filled with water. Worley Junior and I always split our catch and we both had fried fish for dinner at our homes.

There was another way to fish that resulted in catching a lot of fish. One of our Bulgarian boarders, John Lisconoff, took willow saplings and made a basket about five feet long and about two feet in diameter at one end, with all the willow branches coming together at the other end. There was a cone-shaped insert at the large end with a ten-inch opening at the bottom of the cone.

John placed the willows close enough together to hold perch and larger fish but smaller fish could easily escape. When he wanted fish, he waited until the sun went down, then anchored the fish trap between two large rocks in the middle of the river with the open end facing downstream.

Fish swimming upstream would swim into the trap and couldn't get out. The next morning, John would empty the trap into a burlap sack and hide the trap in the brush along the riverbank until he needed it again.

Worley Senior took his sledgehammer and once a week, he would walk along the river's edge and smash the slabs of rock along the shore where many of the fish would swim for cover because the water was so clear. Once he smashed the rock, he would often find two or three rock bass or trout under the piece that fell in the water. In a short while, he had all the fish he needed.

In the woods and forests around every mining town we lived, at least two of the four edible types of mushrooms grew. All that we needed was a pocketknife and a large basket or paper sack. Most people were afraid they were poisonous and would not pick them but we thrived on them. We had creamed mushroom soup, mushrooms fried with onions and peppers, and mushrooms in scrambled eggs with fresh homemade bread made a wonderful lunch. The "oyster white" mushrooms (Galba) were found around poplar wood stumps and logs. Sawdust mushrooms were found around old sawmill dumps, "point-penkias" were found under beechnut trees and regular mushroom were found almost any place it was damp and there were a lot of logs.

When there was a bounty crop of oyster mushrooms, we cut them in strips and put them on a screened window frame in the garden to dry for a couple of days. Then they were placed in paper sacks, the sacks tied shut, and hung in the pantry. When we needed them, all we had to do was soak them in water overnight, drain them, and cook them as usual.

I went to school in a large two-room schoolhouse diagonally across the street from our house. Grades one through six were in one room and the rest of the grades in the other. There were two long church-type benches across the front of the room, called "reciting benches". A large potbellied coal stove heated the room. The stove sat inside a sandbox so that hot coals wouldn't fall on the floor.

Each grade in turn was called up to the benches and did whatever their teacher, Mr. John Perry, wanted them to do—reciting or putting math problems on the blackboard. It was very strict and quite effective. The rest of the classes stayed busy at their desks. There was no walking around or daydreaming.

There were two outhouses behind the school. If you needed to answer "nature's call", you held up your hand with one or two fingers stuck up. In the back of the room, next to the door, was a piece of cardboard hanging on a string. On one side was the word “IN” printed in large letters and on the other side was the word “OUT”. The teacher would glance at the card and if the card read “in”, he would nod yes to the student with their hand up. As you went out, you flipped the card and when you returned, you flipped it again.

There was a small table at the back of the room and a dipper to get a drink with. The other room was used for our weekly assembly. We would sing songs, led by Mr. Perry and Mrs. Jane Booth, the other teacher. I still remember the songs: Nellie Grey, Let the Lower Lights be Burning, Spanish Cavalier, Capital Ship, and America were just a few. At recess, we played games such as “Drop the Handkerchief”, “going in and out the window”, we had races, played marbles and jumped rope.

Then the worst thing that could ever happen did. A bloody strike through Boone, Mingo, and other counties erupted. It became increasingly difficult to earn a living when the coal companies kept thinking of ways to take more and more money out of the miner’s pay envelopes every week.

The word “union” was not a good word to use. The coal companies hired the Baldwin Felps Detective Agency to seek out those people. Some disappeared overnight while others left town in a hurry.

My father and our boarders were all on strike and assembled in a valley on the other side of the mountain from our house. There were thousands of men there, getting ready to march. All were armed with everything from pistols to a 45-70 long rifle of 1896 vintage.

One day as I was playing on the side porch, I heard the drone of a plane—a real rarity in those days. I watched the plane as it flew past. It suddenly went into a long dive, then pulled up and disappeared over the mountains. Then I saw a long object falling from the sky, tumbling end over end, and making a sound like someone hitting a metal pipe with a chain every time it rolled over.

It landed with a thud in the riverbed, just a few hundred yards from our house. My father and some of the other men dug the object out of the mud. It was a twelve-inch diameter cast iron well casing pipe about six feet long. A screw-on reduction cap at one end held a plunger to detonate it. The other end had a four-blade propeller vane on it. Each vane was triangular shaped, made of steel, and was supposed to rotate to keep the bomb's nose down.

I was told later that the coal operators had the bomb made and wanted it to land and explode for "maximum kill". Fins such as those used on regular bombs would have buried the bomb deep in the earth and would not have been as effective as propeller vanes.

I watched from a distance as the bomb was dismantled. It was full of black powder in small bags, and railroad spikes, nuts, and bolts. When it was stood on end, it was taller than any of the men in the picture who helped dig it up.

It was learned later that the WWI surplus plane used to drop the bomb had a "belly rack" to carry the bomb. When the pilot went into the dive to release it, one of the vanes got bent against the fuselage of the plane because the front release worked just a fraction of a second sooner than the rear release. The bent vane was hitting the well casing as it tumbled down, which was the reason we heard the bang and clanging noise.

The strike went on. One day, my father was with a group of fellow coalminers who were looking for thugs hired by the coal company. They came upon a farmhouse in a mountain pass and saw a machinegun set up on a tripod on the porch. It was aimed down the pass. The miners quietly rushed the house, seized the machinegun, and captured two men who were eating lunch in the kitchen. They found even more weapons inside.

The two men were brought to the schoolhouse across the street from our house. Several dozen men milled around the schoolyard and a trial was being held. I climbed up on the fence in our front yard to watch.

It was a while before the two men were brought out and put in a wagon. A group of men followed the wagon as the wagon left town. My father came into the house and Mother asked, “What will they do with them, Yosef?”

Father’s reply was the classic reply. “Who knows?” He said in the Slovakian language. We never saw those men again.

The governor of West Virginia ordered a truce between the miners and the mine operators for a few weeks. When the truce ended, each mine entrance had machine gun emplacements around them. The governor had “sold out” to the operators.

The federal militia was ordered out to patrol and disarm everyone in the area. They came in with guns and marched in front of our house, setting up a camp in the field across the road. As usual, I was on the front fence taking it all in. A nice young man with shiny boots came up to me and asked if I knew where he could find cold drinking water.

I led him around the house to the well on our back porch. Mother came out and the man asked her for water. She said it was good clear water. The man left and a few minutes later, several men with collapsible canvas buckets appeared and hauled away a lot of water.

Every day, the young man, an officer, would speak to me as they left on patrol. One morning, I looked out the front window and all the soldiers were gone.

My brother Steven was born on November 7, 1921, adding to our already large family. Finally, after all the murders and problems, my parents decided it was time to move on. I went to say good-bye to the Worley’s. Junior was a great playmate and I began to cry when we left. I loved that place.

The Diviak children in Jefrey, West Virginia in 1921
Front row: Joseph, John, Steven, and Julia
Back row: Helen, Anna, Mary, and Teresa
(Only one dared to smile.)

My father and the twins, Frank and John, found work at Powel Creek Mine. The four Bulgarian men found work in another mining town where they had some friends.

The rented boxcar was loaded and my father climbed inside. Our family and Frank and John boarded the train for Powel Creek.

Powel Creek was a nice clean place, but the mine ran out of coal in just a few months, so it was time to move again.

This time, we moved to Julian, West Virginia. This mine was located halfway up the mountainside and had a fairly large seam of coal going on for miles at the same level. They used mules and donkeys to pull the coal cars in the mine to the long enclosed chute running down the mountain to the tipple and into the railroad coal cars that pulled underneath it. The whole setup was operated by gravity.

Julian was located on the main railroad line, with a sidetrack to the mines for empty cars to be loaded and another for the full cars waiting to be hauled away. My father and Frank and John, along with the other miners, had invested their savings in the mine with dreams of becoming wealthy. It was a small clean mine, free of sulfur and gas and the prospects looked great.

We were introduced to a new wild fruit that grew all around the mountains there. It was called "paw paw" or "papaw" and is a nice juicy fruit and very delicious when allowed to ripen in the sun on a windowsill. There were also a lot of wild partridges, which fascinated all of us. We would watch the chicks run and hide at a signal from their mother, who would then fake a broken wing to lure the danger away from her chicks.

The freight and passenger trains roared past the houses by the tracks. The engineer would toot the train whistle whenever I waved at him. Soon the engineer started throwing a box of Crackerjacks down into the yard where I stood watching. This went on for a few months, until the engineer saw the dray wagon backed up to our house. He waved his red bandana at me—his final good-bye. It was a sad day for little Joe.

It was a long train ride to Huntington, West Virginia. We had to change trains at one place and take a taxicab to the other station. It was the first automobile ride for the whole family!

After a long two days and one night on trains, we finally made it to Yukon, West Virginia and from there, we walked two miles up a dirt road to Caretta, the town we had left six years before.

CHAPTER FOUR

The Diviak Family's Last Move Together

We passed the new Caretta Shaft Mine and the old Caretta Company Store then the tipple with all the chutes dumping coal into the railroad cars.

A Hungarian family, the David's, had a large room for us to bunk in until the company could give us a temporary house. They were very nice people and they had children. We had a great time together.

Frank and John boarded with another Hungarian family and our furniture was stored temporarily in a small vacant house half a mile away.

We finally got a house near the railroad tracks and spent a day scrubbing it with boiling water and lye soap. We heated the water on a wood fire in the back yard and scrubbed the whole house—the walls, ceilings, and floors—to get the soot and dirt off. We were never bothered with bugs or mice after we moved in.

Even in those days, we did have a few "trashy" families. The mine foreman, our acting mayor, would tell them, "Clean up or move out!" But some were just born that way and never got over it. Later, a company registered nurse made periodic inspections to make sure there were no hogs kept under the house or chickens in a room inside.

In the 1920's, hogs, chickens, and cows were kept in backyard barns or sheds. Later, the coal company sent a nurse and a representative from the main office in Coalwood, a sister mining town owned by the Carter Coal Company, to get rid of all the animals. Cows and hogs were moved outside of town, up the hollow. Chicken houses were built farther up the hillside.

For weeks, the town was covered with smoke from sheds and barns as they were torn down by company employees and burned. Needless to say, the general health of the families improved! Then we were all given typhoid, diphtheria, and smallpox vaccinations, which together practically eliminated illness in the miners and their families.

At the same time, a school dentist made his circuit of each grade school in Caretta and Coalwood. All the work he did was free.

Captain Laird had been the mine foreman at the old Caretta Slope Mine for several years then he was given the shaft mine at Coalwood. Later he became the general manager of all four mines owned by the Carter Coal Company.

Captain Laird was the most well liked mine foreman we had ever had at Old Caretta. When I was a first grader, I would walk home along the railroad tracks, past the tipple. Whenever he saw me, he would come out and talk to me. Often, he would touch my cheek, as though he was brushing off a speck of dirt or a stray hair. When I got home, Mother would ask where I had been because my face was dirty. Captain Laird loved all of us children and he was highly respected because he knew how to handle the men.

Captain Laird was a graduate of Stanford University in California and went overseas in World War I as a captain in an Infantry Company. One Private in his Company was a youngster named Jimmy Carter, son of George L. Carter, the owner of Carter Coal Company as well as a large farm and several small railroads in Carroll County, Virginia.

Captain Laird made Jimmy his orderly and messenger on the battlefront. After the war ended, Jimmy invited Captain Laird to his home in Hillsville, Virginia to meet his parents.

George took an instant liking to this giant of a man and gave him a job as a section foreman in the mines and later, mine foreman. As foreman, Laird made a lot of changes in the mines and he made even more as he was given more responsibility in the Company.

Carter Coal Company had four working mines. The Old Caretta was a slope mine with a four-foot seam of coal that sloped in under the mountain from the valley floor.

The shaft mines at New Caretta and Coalwood each had 650-foot deep shafts that ended in a fairly flat six-foot thick seam of coal. Although a mountain separated Coalwood and Caretta, they both worked the same seam of coal. They weren't joined until just a few years before the seam was worked out.

The fourth seam of coal was in a part of Coalwood called "Six"—six was the number of the mine. It was a few hundred feet up the mountainside and had its own tipple like the other mines, but was a smaller operation. All the coal mined there was bituminous coal, a soft but sparkling coal. There was always a six to twelve inch layer of coal above a soft shale or ash streak between the ceiling and the bituminous coal. It was called "candle coal"; it was very delicate and burned like wood, almost smokeless.

The shale, or "bone" as the miners called it, had to be cast aside on the "gob" side of each tunnel.

A story of how George L. Carter secured all the mineral rights along with the surface land was quite interesting. In the early 1900's (probably between 1901 and 1910), he had stopped for the night at a hotel near the railroad station in Washington, DC, on his way to New York. He was very tired and decided to take a nap. He was awakened by loud voices coming from the room next to his. A locked connecting door separated the two rooms and George listened quietly and learned to his astonishment that the men next door were apparently civil engineers. They were discussing metes and boundaries, bench markers, and such, in an area not yet named, that George was interested in (which would later become Coalwood and Caretta). George quickly copied down all the figures on a map he had along with some notes and returned immediately to Hillsville. He and his own engineers left quickly for the area and secretly bought or leased all the property in the area before the people in New York could act.

My father, who had known George L. personally since 1912 when he first began to work for him in Caretta, told me this story. My father always had an "open door" to visit with George L. any time he wanted, although he rarely did because he knew George was a very busy man.

My sister Josephine was born on February 1, 1924 and we needed more living space. We also wanted to get farther away from the noisy railroad tracks. Our family was now complete—six girls and three boys and all in excellent health.

One day, my father came home happier than usual. "Pack up all the dishes, canned good, clothes, and everything else!" he said. "We move the day after tomorrow. I've hired the Coal Company dray wagon and driver to move us to house number 48 in Poplar Forks." This was about a quarter of a mile from where we lived on "East Street". We had lived in that house in 1913 and 1914 and Helen was born there.

It was one of seven houses with a stucco finish and plaster-on-lath walls. It had a large walk-in closet and pantry. The house was nestled in a row of houses about forty feet apart, set between two mountains. The sun came over the mountain at 10:00 A.M. and set over the other mountain at 3:00 p.m., which made it nice and cool in the summer. It had a separate washhouse and a stove with a hot water compartment for bathing.

The next day, we went to the new house and scrubbed the pine floors until they were almost a light tan color. We moved in the following day. This house was more dust-free and quieter than the other house.

At last, we had a nicer place to live, in a neighborhood with more foreigners, and we shared our holidays in most cases as they did in Europe. We had our chickens, hogs, a cow, and geese. Each of us children had our chores to do. We all instinctively knew to help all we could so our family could survive. We didn't have to be paid to do chores. As one member aptly put it: "If we all did our part, we would all get to eat!"

My job was to carry feed a quarter of a mile to the pigpens, located outside the housing areas, three times a day. I would also drag fallen logs home and saw them up, then split the foot-long pieces into kindling to start fires under the coal. Coal needs a starter and paper won't do it.

The coal stove needed wood every day all year round. Once the fire was going, we added lumps of coal all day, turning the damper down until a higher heat was needed. The fireplaces and "heatrolas" (coal burning heaters), much nicer than "potbellied" stoves in the living room, had to be maintained with wood and coal buckets ready all the time, except in summer.

Water had to be hauled from springs in the sides of small hollows about a quarter mile above the houses in the large hollow. At first, I carried two one-gallon jugs with an "ear" on each side and a cork stoppers. Later, as I grew stronger and learned to stop to rest, I carried two water pails.

We had a garden on each side of the house and one on the hillside. They required a lot of work to prepare. We dug them with a shovel or shading fork, row after row, until it was finished. I learned early not to worry about how hopeless it looked at the beginning of each spring, but to take it one day or evening at a time until it was finished.

Mother Helen Diviak in one of her three gardens. She won cash first prizes three years in a row. Contests were sponsored by the Carter Coal Company. Caretta, West Virginia July, 1933

We didn't have commercial fertilizers back then, so it was my job to get the cow manure from the barns, which were out past the last house in the hollow, and haul it to the gardens.

When I was nine, I wanted a sled that I could use on the slope near the house in winter. I found some old pine flooring stacked up behind the washhouse. Each piece was about an inch thick and three or four inches wide with a groove on one side and a "tongue" on the other.

I got an old saw and cut two runners five feet long, then sawed an angled piece off the end of each one. Now I had my skids. Then I nailed crosspieces to the skids, about two inches apart, from one end to almost the front. The last piece on the front had to extend far enough out on each side to tie a rope to it. I used that sled until Mother "suggested" that it would hold three bushel baskets filled with dry cow manure tied down to the crosspieces.

The first snow, on a Saturday and Sunday, I pulled that sled loaded with cow manure dozens of times to the gardens. I dumped the baskets about three feet apart, in rows. It seemed to me like such a hopeless task! It was like being given the job of emptying a tub full of water with a spoon! But we did have the best vegetables in town. In fact, the Coal Company gave prizes for the best, nicest, and most complete garden in Caretta and Coalwood and Mother won first prize—$100—three years in a row and an alternate prize of $50 each year.

She sold head lettuce, bunches of green onions, radishes, kale, peppers, tomatoes, and more, to the company store. In return, she was given "credit slips" that could be used to buy chicken feed, feed for the cows and pigs, and 100-pound sacks of flour and sugar. The larger quantities were delivered to the house by dray wagon from the company store.

I saw an ad about how good cloverleaf salve was for chapped hands and lips in an old magazine at the store. I ordered a dozen cans to sell at fifty cents each. The cans were about three inches in diameter and an inch high, with a smooth snap-on lid. I went door to door and immediately sold every one of them, so I ordered more of them.

There were prizes for selling a certain number of cans. I think I sold about seventy cans of the sweet smelling Vaseline-type salve before I had enough for a nice red wagon with 10-inch rubber tired wheels. It was just what Mother needed, for me to haul produce to the company store. The wagon held three bushel baskets, tied down. I made the deliveries, pulling my produce-laden wagon approximately three-quarters of a mile to the store, very early in the morning, before school.

I always felt a sense of satisfaction each time the company store bought our produce. After all, we grew those vegetables on company land and used fence posts that we cut on company land, in the woods in the hollow. I felt that at least we were getting a little back for all the charges we paid the coal company to live and work there.

I was a shy, bashful boy and when I went to school, I was always afraid the other kids would "make fun" of me for hauling produce to the company store, until one day in the third grade, a beautiful little classmate, Daisy Baily, came up to me one Friday morning.

"Joseph," she said. "I hope you deliver some more of those good vegetables to the store in the morning. Mother buys them every Saturday morning!"

That did it! No more being afraid of being teased or heckled after that!

Daisy and I were always great friends. It didn't bother her that I was a "foreigner" and lived in "Hunkie Hollow". The blacks there could never say "hungry"; it always came out "hon-grey", so instead of "Hunkie", it became "Honkie" to them. Of course, we had our own name for them as well, but were punished by our parents if we ever used it!

The third group lived in what the kids called "white trash row". If you didn't fall into either of those groups, you were simply "white trash"! But we never fought about it. None of the children were insulted by these names. In fact, when asked where you lived, you replied, "I live in the fifth house in Hunkie Hollow" or "The ninth house on 'poor white trash row'."

If a coalminer had children who were troublemakers or undisciplined, he could be given a "pink slip" and told to move. There were always people waiting for his job as soon as it became available. When you were fired, you couldn't stay in town, nor could you find work in the nearby mining towns. The foremen kept each other informed by telephone. We called it "black listed".

The primary goal of a mine foreman was to produce coal and it didn't matter to him if you were white, black, or anything else, as long as every coal loader produced 16-tons or more of coal, or if you were a timber man, trackman, or any of a dozen other workers who worked inside the mines or assisted the coal loaders to achieve their goals. The foreman was a 'nice guy', regardless of your race or color when you went underground. At the end of the shift, everyone who came out was black, with white teeth, white eyeballs, and red lips.

Our part of Caretta was called Poplar Fork, or "Hunkie Hollow" as the kids called it, where most of the foreigners lived. It was the nicest place because the houses were farther away from the tipple where the large railroad cars were loaded by dumping the smaller mine cars into them.

The tipple had shakers and screens that sorted the coal from coal dust and nut-size pieces to the larger lumps of coal. Each railroad car was loaded with only one size coal, yet there were always five or six cars being loaded at the same time. As you can imagine, it was very dirty in that area.

Poplar Fork Hollow was named for all the poplar trees that grew there between the mountains. It ran for a mile, and then eventually went up to the mountains at the end.

The houses were in a row, about forty feet apart. A dirt road ran along the back and another along the front of the houses. Beyond the front road was a creek that ran all year round and proved to be the best water for washing clothes and watering the gardens. Some even drank it, but generally, drinking water was hauled in buckets from mountain springs just past the last house. The whole area was maybe a quarter of a mile long.

Our house on the right.
A view down the Hollow to a slate dump burning on the side of a mountain.
(1956)

A view up Popular Fork Hollow in Caretta, West Virginia.
(1956)

A ten-foot high rock cliff began at the creek's edge and ran along the creek, up the hollow. On top of the cliff, the steep mountains rose up, covered with trees, mountain laurel, rhododendron, tulip poplar, green oak, chestnut, hickory nut, beechnut, hazelnut, black and light walnut, ironwood, pines, cedars, and more. It was a very beautiful place, summer and winter.

A water hydrant was by each house, but the water was tainted with sulfur. Some people would fill a tub with this water and let it sit for eight or ten hours, then skim off the scum and use the water. Of course, the coal company charged fifty cents a month for that, along with the rent for the houses, plus the coal delivered to each home's coal bin, This coal was used to heat the houses—in open fireplaces, potbellied stoves, and cook stoves. Then there was an electric bill to be paid each month. All these charges were deducted automatically before each miner's money was put in his pay envelope.

The coal loader was the number one position in the mines. Everything and everyone depended on the coal loader producing 16-tons each and every day. But they were also the ones who had to provide their own tools, which they could buy from the company.

They needed a pick, shovel, a large double- or single-crank drill with a breast plate and a six-foot twist bit, a sledgehammer, a flat, long handled dust shovel, a one-man crosscut saw, a canvas bag for dynamite and blasting caps, a miner's lunch pail, a long wooden tamping rod (used for inserting dynamite into a hole), a 100-foot cable, a hotshot, a gas mask canister, a safety helmet and safety-toed shoes.

The coal loader had to purchase blasting caps and dynamite at the powder house every morning before he went to the man trip, which took him under the mountain to the tunnel he was to work in that day. The seam, like jelly between two slices of bread, was 48 inches high. In order to stretch the kinks out of your back, you had to lie on the floor.

Each miner, if he was a coal loader, was given eight or more tags. Each was about an inch and a half in diameter-a little larger than a fifty-cent piece. There was a hole punched out near the edge and a large number stamped on it. This was your number as long as you worked at that mine. My father's number was 56. (Later my number was 356.)

When you went to the entrance of the mine, you gave the lamp man a "check" – your brass tag. In return, he gave you a headlamp with a charged, acid filled battery attached. He hung your tag on a board full of nails, with each man's name and number under a nail.

Once you got your lamp, you would strap it on the rear, wider end of your belt (The belt was cut lower in front to allow easy bending), on your right hip pocket side. On the other hip, you already had a 2-pound gas mask and a clip pin with extra "checks".

This was the coal company's unique way of knowing to charge you eight cents a day for the use of the lamp. (A loaf of bread or a quart of milk was five cents in those days.) A bookkeeper would note all the 'checks" on the nail board after everyone was inside the mine.

At the end of each shift, after everyone had "checked in", if there was a "check" left on the board, they knew that person hadn't made it out, especially if all the lamps weren't in the charging rack. It happened a few times.

The coal loaders, like piece workers in a factory, also had to give a "check" for dynamite and blasting caps. The "check" was placed on a nail board and a number penciled under it. For instance, "3-3" might be under his "check"; that meant he had gotten three sticks of dynamite and three blasting caps, for which he would be charged eight cents each, or forty-eight cents.

It was possible that a coal loader would receive no money for two out of three paychecks. The company was in full control and wanted its money for tools and materials first. Some companies would deduct only a little each month, but with rent, water, electricity, and coal, it was a difficult time for miners who didn't have money saved from previous employment at another coal mine.

The other workers, inside and outside the mines, we called company employees, and received a fixed salary. They were paid every two weeks, as were the coal loaders.

Because of insurance, each coal company had a company doctor. Before you could be employed, even if you had a slip of paper from the mine foreman accepting you, you had to pass a physical. Loss of toes, fingers (but not too many), teeth, and scars was not a problem, but heart problems and poor hearing would be. Once you were hired, you paid insurance out of each paycheck.

If you were married, and you were killed or crippled for life, your family had two weeks to get out of the company house. The state did provide some compensation money for the family, but after six months, that ended.

There were no movie theaters, radios, or televisions, so outside of working nine or ten hours a day, there was no entertainment. This might have contributed to each family having so many children! As one wise old bird put it, "it wasn't until the sixth or seventh child that the parents figured out what was causing it!"

Some of the men played cards, the women sewed or got balls of cut up old fabric and made woven rugs and such. It seemed that everyone stayed quite busy.

The structure called the tipple sat astride the railroad tracks. It was high enough that the huge railroad cars could load up to fifty tons of coal each. Each chute had different size coal for the markets up North. All the railroad tracks were built sloping away from the tipple. Each car was filled by gravity as it passed slowly underneath. The empty cars were uphill and the filled cars downhill on a siding track.

There were several huge locomotives used to collect the full cars and pull them to the town of Iaeger for distribution. The main locomotive was a double-drive monster that could easily pull a hundred and fifty filled cars at a time.

The Diviak home was the fifth house from the end of Poplar Hollow, not a far walk from the little hollow with the springs of the best water around. The house was always nice and cool. We never needed fans. In fact, I had never seen one then.

There was no indoor plumbing. Ex-Captain Laird's training made sure that the outhouses were all lined up, about twenty-five steps along the hillside, doors facing the hillside, lined up in perfect formation. Stepping-stones led to the outhouse. Each outhouse was firmly mounted on a box frame that went down four or five feet into the ground, enclosing the bottom. Each one was painted a dark green to blend into the foliage of the hillside. Needless to say, it was quite a challenge to make a run for the outhouse on a rainy or snowy night!

Each outhouse had three holes cut in the bench inside, properly notched in the front of each hole. As you walked in, the largest hole was to the right; the medium-size one was on the left, and a small one in the center—accommodating all sizes in the family.

When each pit was half full, a new pit was dug next to it, a new box built, and the outhouse was moved to its new foundation. Several bags of lime were spread in the old pit and it was filled in with dirt from the new one. It was very important that the new location be in perfect line with the other outhouses.

It was sort of a status symbol to us kids that we had graduated from a "two-holer" to a "three-holer" when we moved from another part of town! We made certain it was shown to every playmate that came to visit. It sounds crazy, but read on! (Author's note: Some sixty years later, on a guided tour of Cape May, NJ houses built over 300 years ago, with awe viewed the first house to have an indoor flush toilet in America. The house was owned by the Crapper family!)

The "toilet room" was built on the second floor, on the side of a beautiful home. It was roughly six feet long, three feet wide and seven feet high. A down sewer line, hidden in the wall, went to a cesspool in the backyard. The water closet was near the ceiling and had a pull chain to flush it. The "toilet room" was purposely built on the side of the house that could be seen from the street and was painted white, which was not the color of the rest of the house. Of course, it instantly became a status symbol and everyone had to have one. And they were always built on the side of the house—never the back. It also required having a cesspool built in the backyard, which was covered by flowers or grass. So, you see, we weren't too far wrong to be so proud of ours!

Inside each outhouse, you would find a Sears or Montgomery Ward catalogue, tissue saved from packages, and newspaper. Toilet tissue was unheard of in those days.

We did use the water hydrant by the back porch—after all, it did cost us fifty cents a month! We managed to buy a garden hose which we used to water the garden, but the water was rusty, smelled of sulfur, and wasn't fit to drink. On Sunday, we would fill tubs with the water and let it sit overnight. Every Monday morning, we would skim off the rust and other debris and the water was used to wash our clothes with homemade lye soap. The clothes were boiled in large tubs over a fire in the backyard to get them clean. A scrub board and a bluing rinse helped, too. But water from the creek was still the best.

One day when I was a young teenager, I decided to try a homemade cigarette like my father smoked every evening in the backyard under the shade of the peach tree. He used a pipe tobacco called "Five Brothers", rolled up in a cigarette paper that had a glue strip on one edge. I did a great job of rolling and sneaking a couple of matches to light it, went up the hill to the outhouse.

My mother was watering the garden with the garden hose, using a twist sprayer with a fine mist.

I sat down on the toilet seat, lit the cigarette, and noticed the one-inch holes at eye level in the back and around the top on all sides. I was doing just great, between coughs, when suddenly I was hit in the face with a blast of cold water, knocking me against the back wall of the outhouse!

I came out of there fighting mad. I noticed that the nozzle of the garden hose was stuck in one of the eye level holes and there was Mother, standing by the water hydrant by the back of the house. Mother said innocently, "Excuse me! I thought the toilet was on fire!" and went inside the house.

I never smoked in the outhouse after that!

CHAPTER FIVE

School Days and Working in the Mines

In 1922-23, Caretta had a four-room schoolhouse on Tennessee Avenue, which was named after all the families from Tennessee who lived there. And, like Jeffery and so many other towns, the school had no indoor plumbing so a couple of outhouses provided the necessary accommodations.

Right next door, a new elementary school was being built. It was two stories above the basement, which was mostly aboveground. The land it was being built on was the only really flat place in town that had enough room for both the school and a playground. It also had a gym that could also be used as an auditorium, complete with a stage with purple velvet drapes. It was a fantastic place for us kids and looked to us like a palace. Oh yes, it even had indoor plumbing complete with showers for the boys and basketball teams!

The floors were pine and were oiled and wiped every week. The boys playing basketball were granted the privilege of waxing the gym floor by hand, using towels to polish the floor, every evening after supper. We were glad to do it.

The school housed in the new school went from Primary through the sixth grade. Only one class for each grade, and never more than twenty students in each class the entire six years I attended there.

William B. Gray, the caretaker and custodian of the school, was the grandson of a slave from Virginia. We could always go to "Uncle Bill" to get things like the baseballs or basketballs fixed. We all loved "Uncle Bill" and he returned our love by doing anything he could for us. He always had a box filled with assorted buttons, pins, and such. I remember one time he sewed on a very crucial button for me. In those days, zippers were unheard of, and playing tag often resulted in lost buttons.

What was really sweet was the fact that, with a few exceptions, we all helped each other and took care of each other, forming friendships that lasted through high school. We were in school plays together when, to my horror, I actually had to take part in some of the stage plays! Some were in front of our parents, held at the end of the school year, or at Christmas, or occasionally in the fall.

One day, at the end of the school day (which was 4:00 P.M. in those days), my third grade teacher said to me, "Joseph, I will walk home with you today. I want to ask your Mother to sew a black costume for me for you to wear in the school play."

Horror of horrors! I was beside myself. Everyone will think I did something wrong! Or worse, the malcontents will call me "Teacher's Pet"! What will Mother think? My little mind was racing but there was no escaping it.

My teacher, Mrs. Mary B. Tanning, walked home with me and met my mother. When she told Mother what a good boy I was in school, I began to relax a little. Then Mrs. Tanning got out the material and a pattern and she and Mother discussed the costume—Mrs. Tanning in her perfect English and Mother with her broken English, but they had no problem communicating.

Before long, Dad came home and I began to smell cabbage cooking. Cabbage leaves stuffed with rice, ground veal and pork, a lot of chopped cabbage, cut up onions and diced tomatoes, all steaming in one huge kettle. I knew we always had stewed beets and a cucumber salad and fresh homemade bread with this meal.

I really panicked when my teacher was invited to have supper with us! It was sheer torture! But everything went well. My brothers, sisters, and I were on our best behavior, eating much slower than usual. I breathed a sigh of relief when Mrs. Tanning accepted a second helping of stuffed cabbage! We kids called it "pigs in blankets".

When Monday morning came around, I had another worry. I was in great fear of what Mrs. Tanning would say about having supper with us on Friday evening.

Class started after the opening prayer. (I never met an atheist in the coal mining towns, inside or out, so there were never any objections to prayers in school. We figured that we could use all the help we could get to help our fathers in the mines and bring them safely home after each long shift.)

As class began, Mrs. Tanning told everyone how much she had enjoyed the supper and how great the food was. She told the class how she had met my parents and brothers and sisters. Boy, I was Number One in class that day and no one made fun of me again that year. Mary B. was a kind woman; she knew how to use the "paddle" but she was firm and fair.

The first of June, four of us boys who played together all the time would go up the hollow and dam up the creek with logs, rocks and clay, then we would swim in the cold, clear spring water each day—without swim suits. Our legs and arms would turn blue from the icy water and whatever physical assets we might have been blessed with sought a warmer area in our bodies, retreating as far as possible. We were fairly safe from exposing ourselves and no one ever came up the hollow during the day.

However, there was one old, skeleton-like Hungarian lady. She had a hooked beak that passed for a nose and she loved to meddle in things not related to her own home.

One day, she told my mother that she had seen me and my three buddies swimming nude up the hollow. She said that she had gone up to the spring for a bucket of water and heard us yelling at each other, so she went around the bend and saw us.

"You could not see who it was from that distance." Mother said.

She replied, "Oh, I walked closer and watched them for a while and saw it was the four boys who play together."

"I will talk to Joseph about it." Mother told her. "Don't worry." Later she and my father discussed it and I heard Father say that he had done the same thing when he was a boy in Europe and he hoped our pool was big enough for him! "That cold water would kill you!" Mother replied.

Since we couldn't afford swim suits and all our old trousers were used to do chores in, we were told to be more careful and that we should take turns watching. Mrs. K did sneak up a few times after that, but one of us would ask her what she was looking for so she stopped.

The elementary school was governed by a series of tough principals. It seemed there was a different one each year. They were good men and certainly ran a tight ship. We didn't have a local school board or PTA, so the principal could devote his time and energy to the school, not making reports that no one ever read but had to be kept. What a sweet setup. The County Superintendent in Welch was in charge. Coalwood was the headquarters for the four Carter Coal Company offices and all the decisions were made there.

We all owe "the wheel" that thought of having a dentist there for the families. But he was not really affordable, even though a filling was only a dollar, especially if there were a lot of children in each family.

Captain Laird, the mine foreman, decided that we should have a portable dental chair and foot operated dental drill in each school. Ours was in a small room in the basement. Everyone from the first grade to the sixth had to visit the dentist once a year, unless they had a note from their own family dentist in the cities.

Dr. Hale was very good with children and all of us, one at a time, would take our turn. On weekends, he was generally busy with adults in his office in Coalwood. He lived in the boarding house near the company store but some weekends, he would visit his wife and children at their farm in Virginia. He raised chinchillas by the hundreds for their fur and had employees who did the work there.

The only problem I had was fantastic fear I had of the dental drill and the forceps he used. As a kid, those drills sounded like thunder and our heads would buzz for a couple of hours afterward. The Novocain was terrible and going back to class after having a tooth filled or removed was almost a religious experience about to happen!

I learned to shut it out of my mind all summer, but the day after Labor Day, the first day of school, the fear would begin to build again. In retrospect, it was a great program and helped us retain our original teeth with very few replacements. They always gave us a toothbrush and toothpaste after each visit, with a stern warning to brush twice a day. The company paid for all the dental visits. In fact, he needed the work and received a salary each month regardless of the number of patients he had.

Teachers were paid between sixty-five and seventy-five dollars a month and came from all across Virginia, with a few from Kentucky and Ohio.

Every year at Easter, the first through third grades would have an Easter egg hunt on the grassy part of the schoolyard and along the creek bank. It wasn't until years later, when I visited one of my old teachers, that I found out that the teachers had paid for the dozens and dozens of eggs and the dye, as well as the Christmas decorations and little gifts they made for each of us, out of their own meager paychecks. I was very upset because I had never given it a thought. Although we did thank all our teachers, somehow we thought the McDowell County School District Office for the Big Creek District, which covered our schools, paid for all those things.

I remember many a boy and girl receiving pencils, tablets, etc. from the teachers. Again, they paid for it all. They were truly dedicated people and each one played an important part in my life.

My fourth grade teacher was Miss Lois Sparks. She was beautiful, a sweet, small and caring person. At our young age, every boy in the class fell in love with her. The first day, she handed out a mimeographed sheet to each student. Ten word groups covered the page. For homework, we had to learn to spell 10 words each day and know what they meant. When we finished that page, we were given another. At the end of the week, we had a written spelling test—words taken from those lists. Oh yes, the boys did better than the girls! I know we tried harder!

Miss Sparks also had mimeographed pages of the alphabet and we had to do them according to the page and no printing was permitted. Of course, we had started printing and writing in the Primer grade and each year thereafter.

We did long division and percentages the same way. I really began to see why I needed school. I wanted to learn all I could so I could leave that coal mining country! She read books to us and had us read, too. Miss Sparks did a magnificent job with us. We never had any problems in the fifth or sixth grades. We were ready!

At the close of school, on the last day of the year, we received our report cards. We stood holding hands in a circle around the room with Miss Sparks and sang, "Goodbye ole pal, it's time to leave you now..." Many of us had tears in our eyes because we knew that Miss Sparks had accepted a position in another school for more pay and that she wouldn't be back next year. Although we were happy for her, we would all miss her.

Miss Sparks hugged each student as they filed out the door. I, as usual, was very shy and bashful, and pretended to be checking my desk until I heard Miss Sparks say, "Come on, Joseph! You're next and the last!" As she hugged me, she said, "If I ever have a son, I hope he will be just like you." Then she kissed me on the cheek, and said, "Don't ever give up and I will always be proud of you." I thanked her and stumbled through the door with tears in my eyes.

Much to my surprise, many of my classmates were standing outside the door and they began clapping. I was told later that I was red as a beet. Life was tough back then, but it did have its benefits!

During the 4th, 5th, and 6th grades, we were eligible to compete for the basketball team. We had that beautiful gym and had physical exercise there from the first grade on. I went out in the fourth grade and made the team in the fifth grade as a center. I was taller than most and lean and fast.

We had run track (100 yard and 220), vaulting, high jump, broad jump, etc., so we were in great shape. Summers were spent working in the garden and doing chores at home so there wasn't any of that so-called "baby fat" on any of us boys.

We played Bartley, English, and War elementary schools and had great winning seasons. Our coaches were usually the principals of the schools and the referees were usually some of the teachers. It was really fun competing in basketball and at the field meets in War, West Virginia on the field that later became the Big Creek High School football field.

One day in early spring, while I was in the sixth grade, I was practicing pole-vaulting at noon. The pit was a large hole filled with sawdust and our pole was an inch and a half thick and about eight feet long. I was trying to clear six foot the second time when the top of the pole snapped just as I was horizontal over the crossbar. I landed on my back with half my body in the pit and the other half on the hard ground.

I couldn't breathe and kept gasping for air. Mr. Naugle, our coach and principal, wouldn't let anyone touch me. He sent a student over to the company doctor's office, about a quarter of a mile away. I managed to roll over onto my right side and began to catch my breath but I was in a lot of pain. My classmate and friend, Daisy Bailey, came over and knelt beside me. She wiped my forehead with her handkerchief and whispered over and over, "Be still, Joseph. You will be all right."

It seemed like a long time passed before the doctor arrived but in reality, it had only been a few minutes. Dr. Clarke checked me over while I lay there then got me to my feet. He sent me home with some white powder in an envelope. I walked the half-mile home with many rest stops.

When I got home, I took the white powder in a glass of water. The next two days were painful but I still did all my chores—I just rested a lot. I was back in school on Monday but had to give up field and track that spring. My chest hurt whenever I ran or jumped. By fall, I was fine.

After our graduation from elementary school in Caretta and our little graduation party, we all got together to sing songs and say goodbye. Daisy came up to me and told me that her father had been transferred to Fairmont, in the northern part of the state, and she wouldn't be with us in Coalwood Junior High that fall. This was sad news for both of us. Daisy stretched on her tiptoes to her fullest height and kissed my cheek. I never saw her again.

Jimmy Dean and I were great pals. For a year, we had been hitching rides to Coalwood one Friday night a month to attend Boy Scout meetings. The meetings were held in a room that had once been a classroom. We had a great troop of thirty-two boys, a young assistant Scoutmaster and a Scoutmaster. All the boys in the troop went to Coalwood Junior High, so when I went there in the fall, I already had a lot of friends, which helped a lot.

We couldn't afford a Scout uniform for me so, like many of the others, I wore Levi's and blue work shirts. Dues were a quarter a month. I bought an old Scout Manual from an ex-Scout for a quarter, so I was able to make it from Tenderfoot to Second Class and finally, First Class. I loved the Scouts and learned a great deal doing the merit badges. That is where I became interested in Morse code.

We used a "wig-wag" flag, then flashlights at night. We also made a telegraph—a door buzzer, a dry-cell battery, and a homemade telegraph key. Little did I know then that this would eventually lead to radio and a lifetime in electronics?

Coalwood was a 30-35 minute school bus ride from the company store in Caretta. The road wound up through a mountain pass, then downhill to Coalwood Junior High. The pavement was narrow with five or six dangerous curves and we had several close calls, but our bus driver, Jim English, did a great job.

My first few days at Coalwood were an eye-opener! I knew I had to study very hard to make it. We had a different teacher for every subject and it was a challenge every fifty minutes to make the classes in other rooms upstairs and downstairs. One class, Industrial Arts (manual training), was held in the basement.

The principal, Charles L. Righter, told us the first day exactly what was expected of us. Mr. Righter made sure we all did our best in the classrooms and physical education. He became my role model—I wanted to be just like him.

Mr. Righter was our Principal, our Coach, our Science teacher, and taught physical education (P.E.). The first P.E. class, he took all of us boys into the gym and showed us a rope hanging from a steel beam at one end of the gym. "Before the end of this school year," he said, "you will be climbing that rope with your hands only, not using your feet! You will then touch the beam and come back down slowly, hand over hand."

I was stunned. I climbed trees and grapevines in the mountains and hollows in Caretta, but we always used our feet. This seemed impossible.

Five days a week, we did calisthenics for an hour each day. We went from one exercise to the next. In between, we practiced basketball. I began to see a wonderful change in my physical make-up. It was nice to feel strong and healthy.

Then came the running and tumbling on the mats and we all became more self-confident. At the end of the school year, in the seventh grade, we did climb that rope and touch the steel beam, just like Mr. Righter had promised we would!

I went out for football. Our History teacher and the Industrial Arts teacher were our coaches. Mr. Hawkes was a fine man and we all did our best for him. He was the seventh grade—the "scrub team"—coach. Mr. Righter coached the varsity team. One day, Mr. Righter's pocket watch disappeared. He had laid it on the window ledge. One of the new football players had taken it and was terrified to admit it.

The next morning, I was called into Mr. Righter's office and a Mr. Dorfman began to question me about the gold watch. He believed that I had taken it. I didn't know Mr. Dorfman, but had always disliked his looks. A lot of the kids said he looked sneaky or like he was out to get someone.

I answered his questions but they gradually became more insulting and intimidating. Finally, Mr. Righter came in and said, "Joe, you can go now, but, if you can, help us find the watch."

I went to Jack Long, the football player I knew had taken the watch, and told him that I knew he was the only one in the area where we suited up when the watch was taken. I told him that if he had the watch, to please give it to me if he was too afraid to return it himself. He silently handed me the watch.

I kept it overnight. Early the next morning, our bus arrived before the others. I went directly to Mr. Righter's office and laid the watch on his desk and left. No more was said about that watch until some fifty years later—this is explained in a later chapter.

After I suffered an eye injury in football practice, I dropped out on the advice of the company doctor in Coalwood. I did heal enough to play basketball as a sub in the seventh grade, making the team in the eighth and ninth grades.

Coach Righter called us his Notre Dame team because we all had foreign-sounding names. We loved it and worshipped him and in turn, we made him proud of us by winning over seventy-five percent of our games against other junior high schools.

In our Science classes, Mr. Righter was a true professional. He always wore a suit and tie, though on very hot days, he would remove the coat. He was always clean-shaven and his slacks and shirts were always neatly pressed. His ties, socks, and shoes always seemed "to match"—nowadays, this is called 'color coordinating". He was a wonderful teacher and I wanted to be just like him. He became my role model outside the mines. My father also had a great influence on me at that time, although he rarely dressed up because he worked hard in the mines.

One Saturday morning, I set up the ironing board and put two iron soleplates on the kitchen cook stove to heat. I used the snap on handle to press creases in my two pair of blue denim slacks. Mother always ironed our shirts but I wanted to redo them after wearing them.

My sisters began to tease me about girlfriends and such, not knowing that I was trying to follow Mr. Righter's example of always dressing neatly and having a positive attitude. Of course, the girls at school did notice and commented about it a few times, but I had one goal—my dream—and that was to get out of there "when I got big."

Junior High was the turning point of my life. I knew I could make my dream come true if I learned all I could. I read all the required books and dozens of others as well. I just had to know how things worked and why. Like Mr. Righter, I helped everyone I could with their problems in school, and at home in the hollow, too. In retrospect, I remember how all the teachers looked, everyone so neat and trim.

There were no beards covering their faces as if they had something to hide, no crazy mustaches or fuzzy faces, no baggy slacks, no unshined shoes, no too-short skirts or long clanking fingernails on the chalkboard. Most of all, there was nothing there to distract us from whatever the teachers were teaching us. Students were expected to be clean and neatly dressed with their shoes shined, and ready to learn.

In the ninth grade, our last year at Coalwood, we had a school play in December. The play was called "The Great Pumpkin Case" and I was to be the judge in the play. A pumpkin had grown on another farmer's property and not where it should have been. A cow from a third farm broke through a fence and was eating the pumpkin vine. The pumpkin was dislodged and rolled down the hillside, shattering on a tree stump.

It was a huge pumpkin and could have won a prize at the County Fair. The cow ate a good bit of the pumpkin, which upset the first and second farmers. The farmer who had planted the pumpkin sued the cow's owner and that's when all the fun began in this 45-50 minute play!

My Mother and my sister, Theresa, attended the play that night. It was a very humorous play and everyone enjoyed it. Towards the end of the play, my mother leaned over and asked Theresa, "When is Joseph coming on the stage?"

Theresa laughed, “Mom, Joseph is the judge on the bench behind the table on the platform!”

“I don’t believe it!” Mother said. I had on a wig and painted on wrinkles—my own mother didn’t recognize me! It was a lot of fun.

We had a wonderful graduation exercise, in May 1932, in the evening, held in the auditorium/gymnasium at Coalwood. Our guest speaker was Congressman Key. Our class of ninth graders sat in front, center stage, with the speaker and faculty on the stage. The boys all wore suits and ties and the girls wore nice gowns. Most of us boys also wore a vest and a “dollar watch” on a chain, as men did in those days.

After Congressman Key delivered his inspirational message: “You can be president if you want to lead enough. It’s up to you!” we were each given our diploma and dismissed.

Congressman Key asked to speak to all the boys alone. He had a stern look on his face as he said, “I didn’t mind you boys looking at your watches every few minutes, but when you began to shake them to see if they were running, that really hurt me deeply!”

He looked around at the sad expressions on our faces and broke out in a booming laugh. “My class in junior high did something very similar.” He said. “Don’t give up, boys. Keep trying and you will be fine!” He shook hands with every one of us.

I applied for a job in the old Caretta Slope Mine on June 1st and was given a slip of paper to see the company doctor. I passed the physical easily and was assigned to be a trackman’s helper. All I needed was a miner’s lunch pail, a hard hat, safety-toed shoes (“1000-pound capability”), a wide miner’s belt, a gas mask to be worn on the belt, and a pair of gloves. The company store had everything I needed.

I showed up for work at the mine entrance before 6:00 A.M. to catch the man trip underground a couple of miles beneath the mountains. The company clerk gave me five or six brass "checks" with the number 356 stamped on them. I turned in one when I drew a headlamp and belt storage battery so I would be sure to pay the eight cents for its use that shift. If I didn't turn in my lamp at the end of the shift and retrieve the "check", they would come looking for me in the mines. That happened a lot to others over the years.

There were no surprises in the mine back at the working "faces" except that the seam was only 48 inches high. You walked doubled over or sideways like a crab. You had to lie down to stretch!

The trackman was a black man five-feet tall and almost as wide but he was all muscle and no fat. We hit if off instantly and became good friends. Mr. Johnson was a hard worker and made sure I was safe. He taught me when to run or not to run by the sound of the cracking in the roof.

The rails came in 20-foot lengths, with the short rails—the jumpers—shorter and of different lengths. As a coal loader cleaned out his cut of 16-tons, there would be at least a six-foot gap from the end of the rails to the face (the end of the tunnel he was mining). We would put on 6-foot jumpers with connecting fishplates, to the end of the track, and four bridles (steel ties with swivel lugs to latch on to the rail flanges). Then the empty cars could be loaded.

We had 12 or 16 coal loaders in each section, so when we weren't laying jumpers, we were working on the main line, adding rails and wooden ties with spikes, or adding switches along the line for the new tunnels being dug.

Mr. Johnson and I looked out for each other, watching for falling rock from the ceiling, danger boards for warnings during blasting, and the trolley lines on the main entry. These trolleys had 250 DC lines with plenty of amperage. The little gathering motors (locomotives) had a wind-and-unwind cable driven drum that the motorman could use to drop a hook over the trolley wire and go up to 200 feet before he ran out of electrical cable. It was a fast operating assembly and kept the arcing down on the main line where the ventilation was better to keep the methane gas (ch^4) from forming.

In the working areas of coalmines, there was a lot of gas, so precautions had to be taken so as not to cause a mine explosion.

A trackman's helper earned $2.24 for an eight-hour shift, but many times, we stayed in another hour, waiting for all the mined coal to be dumped into the tipple outside. Some mine cars and motor (trains) would bring us out. That was the reason for a strike later on—for "portal to portal" pay, which meant that we would be paid overtime for every hour over the eight hours we were in the mines. I never benefited from that because it came long after I left the coalfields.

In 1930, the Depression hit the coalfields and the Carter Coal Company mines were managing to work only one or two days a week. Some of the workers had to be laid off but not the ones who had built up high charge bills at the company store! They would be the last to go.

Our family was fortunate that we had three gardens and a lot of canned goods, put up in Mason and Ball canning jars with the old-style galvanized lids and rubber gaskets. We gathered all kinds of wild berries for jelly, mushrooms, and nuts. We always had chickens and eggs, three hogs to butcher in the winter, a smokehouse to preserve the meat, cracklins in a 50-pound lard can plus hog lard that made great gravy to put on homemade bread. We learned to "make do", a phrase used by the mountaineers in the area.

Before I went to work in the mines, I had cleared off a plot of ground covered by cuttings from timber that had been harvested up the hollow out of town, with the permission of the mine foreman. I made a brush fence around the area, which was about the size of half a football field. After the rocks were stacked, I didn't need to dig the soil. I planted corn, tomatoes, cucumbers (around the huge rock pile), and turnips. After the corn grew to about 24 inches, I planted Kentucky Wandering beans in each hill of corn. That way I didn't need bean poles. I had great success with my garden and I planted it for several years. We used the cornstalks for the milk cow and the corn for the hogs.

During the height of the Depression, my father and I worked on a dirt road being built that went up and through a mountain pass between Caretta and Big Creek. It was only a path for people and horse riders and the eight or nine of us working on this WPA project worked with picks and shovels to widen the path so wagons and cars could use it.

The bootleggers loved it! We were paid with credit slips to be used at the company store.

The squirrels, rabbits, opossums, pheasants, groundhogs, and fish almost all vanished during the Depression. It seemed that everyone had a 22-rifle or a shotgun. The natives combed the hills for dandelion greens, poke leaves for salads, berries, and such. Some of the people sat in the shade on their porches and laughed at the "dumb foreigners" working in their gardens in the hot sun, but in the early fall, when their money ran out and they were hungry, it was a different story! We did a lot of sharing but I was often reminded of the old fable about the ants and the grasshopper: the grasshopper made fun of the ants in the summer but came to them for food when there was snow on the ground. Somehow, the story wasn't as funny then. A lot of the natives (people whose ancestors lived in these mountains long before the coal mines) had an attitude that seemed to be: "Live it up today while you have it and the hell with tomorrow!"

My father had a beautiful grey three-piece suit that was too big for him so he gave it to a native just before Christmas, which made the man, Liege Ball, very happy. He wore it on Christmas Day and looked great! But the day after, I was on my way to the grocery store and passed Liege's house. There he was, under his Model T car, working on it, and wearing the suit, coat and all! It was a cold day. The suit was ruined but I kept quiet about it. But later, my father saw Liege in the suit and couldn't believe it but he didn't say a word. Leige never received anything from our family after that. We had good reason for calling people like that "poor white trash"!

After the banks closed, and my father's money was in one of them, we still managed to survive without hunger by working in the gardens. A few months later, when the banks reopened, my father, brother-in-law, Fabian (Steve), and I went to Welch to buy a Rockney Six Studebaker. Dad paid cash for the car and I learned to drive it. I had my driver's license at 15 (and still have it).

The 'around the mountain' roads from War to Berwyn, West Virginia, and to Caretta, where I learned to drive at age fifteen in 1930.
Note the three-foot seam of coal at road level on the lower left.
(Picture 1956)

I had to do most of the driving because my father had a stiffness in his right leg and it bothered him when he drove. It was up to me to drive neighbors (free) to the hospital in Welch. I drove Mother and Father to dances in Gary, West Virginia, where they held a couple of dances a year with a live orchestra. I drove to Virginia to buy pigs or a cow or geese or whatever, which the farmer would deliver to our house the following week. We did carry a lot of produce to the Caretta store and another large store in War, W.VA, so the car paid for itself.

We visited friends occasionally, and had a few parties and outings away from Caretta, too. But I never used the car on a date. In fact, I never had a date the whole twenty years I lived there! I didn't want to get involved with anyone and end up married and having children in a coal town. I knew if that happened, I would be condemned to hard labor the rest of my life. It was a battle of self-denial for me. Like most young men, my genes craved excitement and it was a constant battle to keep them under control.

Our neighbors kept asking my parent about my not dating. It seemed that they and some of the girls in my classes thought there was something wrong with me! My parents told the neighbors, "Joseph does not want to stay here. He is leaving as soon as he gets through school. It would be hard for him to find a job outside the mines with a family to worry about and to support."

We were great friends with a Greek family who lived a short distance from us. Their son, George Palus, who was 18 years old, worked in the mines and was taking a correspondence course in Radio Operators and Repairs from the National Radio Institute in Washington, DC. He had heard that I had built a crystal set that worked nicely and then a two-tube battery powered radio receiver from parts ordered through Allied Radio in Chicago.

He invited me over one evening and I helped all I could to solve problems that his instructions didn't cover. I started with his first lesson at home and answered the ten questions at the end of the booklet. Then George would grade them with the correct answers from NRI. I stayed three or four lessons behind him. There were about thirty booklets and then the components to build a three-tube radio with a power supply.

I read all the books available in the Coalwood Library, which had been a high school before I went there. George and I met once a week until we finished the course. It was a fantastic help to me in later years and it helped George get out of the coalmines, which is a story in a later chapter.

The school bus picked us up at the company store at 8:00 am, after it had returned from picking up the tenth, eleventh, and twelfth grade students in Coalwood. We were taken to Big Creek High School at the far end of War, W.VA, at the edge of town, on the Dry Fork River. Students from Berwind, Cucumber, Squire, Newhall and Bishop from one direction joined us there. Another bus brought students from Raysal, English, Bartley, and Yukon in the other direction. That, plus students from War and Warrior Mines, made the consolidated school quite a collection of students!

My class of 1935 was the first class to spend a full three years there. It was a nice, beautiful building built on the only flat area in the Big Creek District large enough for a lighted football field in front of it. It was an ideal place and was centrally located in the district.

Big Creek High School in War, West Virginia
Built on the only flat place in the Big Creek District—just large enough for the school and the football field in front of it on the banks of Dry Fork River. It gave me the tools of learning I needed by having the best teachers, who were experts in getting a lot of knowledge into our hard heads!

Fred Vogelsong and I became friends from the seventh grade on. His father was the telegraph operator at the Coalwood Railroad Station and Fred was an only child and was quite shy. He'd had tuberculosis when he was younger and, although he had recovered, he wasn't able to play sports, nor was he drafted during WWII. We seemed to hit it off as friends.

We brought our lunches in brown bags and ate under a large sycamore tree. There was a large, deep, clear hole of water and we watched fish swim back and forth, mostly perch, bass, and brook trout.

I learned early to cover the smell of sardines or homemade garlic sausage in my sandwiches with a sliver of "Sen-Sen". It was small pieces of spice the size of a pencil lead, purchased from the drug store. A ten-cent packet would last weeks at one a day after lunch.

I had to drop out of football but practiced basketball during physical education. I worked four or five nights in the Caretta Shaft Mine each week—in at six and out at three or three-thirty in the morning. Then I'd walk the mile home and go to bed.

We had an excellent bathhouse at the shaft mine, with lockers. It was similar to those in the Army with no privacy in the showers. I was usually in bed by 4:00 A.M. and up again at seven for breakfast, then to the company store to catch the school bus by 8:00 A.M. .

I wrote a lot of notes on three-by-five cards I made from heavy paper from the dime store, mostly dates in history, algebra and plane geometry formulas, verbs and adverbs, etc. At midnight, we had a thirty-minute "lunch break". The rest of the men would eat quickly, then turn out their lamps and take a nap. I stayed awake, studying the notes I kept in my shirt pocket. I watched the time on my "dollar watch" and woke the others when it was time to return to work.

The next morning on the bus, I could hardly read the notes because they were so smudged with coal dust, but it worked. I never failed a subject.

I carried a full load of 16 credits for college curriculum. I had four electives, one each year: Industrial Arts, commercial English, writing and typing, and business math. I wanted to be ready for anything out there in life.

We had the best teachers. They really cared about us and helped us in every way possible. I competed for jobs in exams, and even against college graduates, I won.

I received my pay envelope every two weeks and gave all but fifty cents to Mother to help with the expense of schoolbooks, clothes, and such. Julia, John, Steven, and Josephine were younger than I and were following in my footsteps. The older sisters, Theresa, Mary, Anna and Helen left school in junior high. Theresa and Mary did domestic work for others. Anna did that too for a while but left at age

17 with Helen, who was 15, for Perth Amboy, NJ. Helen became a seamstress for a large department store in New York City and Anna a foreman in a cigar factory. All were married later but Anna and Helen continued to work until retirement.

It was up to me to be the first of Helen and Joseph's children to graduate from high school, followed by the other four children. I was kept very busy but still found time to build a single tube transmitter and finally a two-tube, 1000 kilocycle transmitter while in the ninth grade. I salvaged a crank-up phonograph and an electrical record pick-up from the garbage dump in Coalwood.

I had scrounged enough scrap lumber from the sawmill up the hollow to build a 6-foot by 6-foot square radio shack on the hillside. The boss at the sawmill let me have the wood free when he found out what I wanted to do with it.

The radio shack had a sloped roof, a floor and vertical strips that covered the cracks in the walls. It got cold at night up there, but I needed a place to work and "keep Joe's radio junk" out of my mother's way. I used a large flattop antenna made from wire salvaged from the motor barn junkyard at the Company's repair shop. The word soon got out that "Joe was playing records on 1000 kc on the radio dial." The neighbors sent some records up for me to play and we had some that Dad loved to hear. Another kid, Carmen Marino, who lived about a quarter of a mile from us, brought some records to play and did the announcing through a carbon microphone I got from Allied Radio in Chicago.

Carmen wanted to be a movie actor because he felt he was so good looking. He even had pictures taken and sent them to various movie studios in Hollywood! Unfortunately, Carmen had the worst nasally voice I have ever heard on radio. People kept telling me to keep "that big-mouthed guy" off the radio!

Finally, the broadcasting came to a halt. One day, the mine foreman stopped my father as he was passing the Company office on his way home. "Mr. Joe," the foreman said, "I admire and like what your son is doing on the radio and trying to learn enough to make it outside. But I had a call from the general manager of the Carter Coal Company in Coalwood. He said that a Federal Radio Commission inspector had received several complaints from Caretta and he is coming to check it out. Tell Little Joe to shut his radio station down for a while!"

I took the radio apart and returned all the phonograph records to their owners. Some of my listeners were quite upset and tried to find out who had complained. It wasn't until a couple of years later that one person did find out who it was. The two culprits were bragging about writing to the Federal Radio Commission. It turned out that they didn't even own a radio! They were just jealous that a "foreigner" was smarter than they were! That seemed to be an ongoing problem back then. If we made good grades in school (and we did), we were called "teacher's pets". If our fathers did well on their jobs and made history by being the best coal loaders the Carter Coal Company ever had (my father did, three years in a row!), he was labeled as the "foreman's favorite". It was said he got the best tunnels to work, which was a lie: coal loaders, my father included, generally worked in the same tunnel for years.

In some mines (although not in Caretta or Coalwood), mules and donkeys were used in the mines and were kept in some mines, like the shaft mines, for a couple of years. There were barns for them with feed and water. The dung was loaded in slate cars and dumped on the slate dump. Slate is a type of rock that comes from the roof or floor of the mines. It has to be moved out to keep the tunnel clear of obstructions to ensure smooth airflow from the giant air turbines. These turbines have a spiral set of vanes that 'screw" the air into the airshaft, forcing it to flow through the miles and miles of tunnels. Eventually, the air comes up the shaft that the men go in on "the cage" (the elevator).

Water and sulfur combined generate sulfuric acid, which actually eats the hooves off the animals. Eventually, the animals had to be brought out of the mine and put out to pasture. Dark glasses were strapped over their eyes for weeks until their eyes could get used to the sun.

We had rats in the slope mines, which had an open entrance where they could easily gain entrance. Most of the rats were white or spotted. They were not harmed by the miners; instead the miners took notice when the rats scurried out of a place because the rats could hear the roof creaking, signaling that it was going to cave in, long before the miners' ears picked up the sounds. The miners weren't far behind the rats getting out of the tunnel! The miners even shared their lunches with the rats. One day, my father was sitting on a large lump of coal in a small side tunnel (called a "break through", between two tunnels but with no tracks), eating his sandwich. A large rat came up and sat in front of him. The rat had a mangled left front paw that was barely attached.

My father pulled a thread out of the sleeve of his blue work shirt and tied the leg off just above the wound. He took a small piece of burlap, used to direct airflow in the working face of a tunnel, and laid the rat on it, out of the airflow. Dad placed a bit of meat and bread and a small jar lid with some coffee (cream and sugar, too!) in it beside the food.

The rat wasn't seen for a few days but the food was left there each day. Finally, one day the rat showed up. He sat with his front paws up so Dad could see that his left leg had healed, minus the paw. Every day after that, my father and the rat continued to share lunch and 'conversation".

A year later, my father was moved to a new tunnel in a very dangerous place, requiring the most experienced coal loader, one who could load more than 16 to 18 tons of coal a day. Dad worried about his rat, but a few days later, a coal loader told Dad that his rat had finally adopted him and had learned to eat “fatback” and “cornbread” and was doing great! No more smoked sausages and pork chops though!

CHAPTER SIX

United Mine Workers Union and School

In 1933, after beginning to recover from the Depression, the coal miners were given permission to organize a union, under various laws. President Franklin D. Roosevelt insisted that this law be passed, and it was.

We would finally be free from the paid thugs and detectives hired by the coal operators to keep the unions out of Southern West Virginia coalmines.

On Friday evening, a large group of miners met with a representative of the United Mine Workers Union in the church building. The building was used by the Afro-Americans for worship and was company property. After much debate, we elected a staff of officers. John Kane, a former Welsh coalminer, was elected President. Bob Miller, an Afro-American, was elected Vice President. George Palus was elected Financial Secretary, and I was elected Recording Secretary. Henry Armstrong, another Afro-American, was named Sergeant of Arms and he selected the code words that we would use to gain access to our monthly meetings.

Our first contract with the coal company was for higher pay and a Union check weigh man to be present along with the company-paid weigh man as each car of coal was dumped onto the tipple conveyor outside the mine. Old coal loaders like my father knew they were being given light or short weights on each car they loaded. Each coal loader had his round metal tag with his number stamped on it that he snapped on to a large safety pin bolted to the inside of each car.

As soon as the car was dumped, his tag was removed and he was credited with however many pounds the car weighed. Five or six cars would haul 16 to 18 tons a day. There was a State mine inspector who periodically checked the mines and the scales, but the Company knew when he was coming and passed the inspection every time.

When we hired a check weigh man who was being paid by the Union, suddenly each car gained three or four hundred pounds! This proved that that Coal Company was stealing from the coal loaders.

Assistant trackmen and assistant timbermen went from earning $2.24 a shift to $3.56 a shift. Coal loaders went from fifty cents a ton to seventy-five cents a ton.

We kept fighting for extra pay for coal loaders who had to work in water in some of the tunnels, called "water yardage", and to do away with Company script that was worth only seventy-five cents on a dollar on Company property and was not accepted anyplace else.

Then there was unfair firing of coalminers by foremen, who refused to listen to the grievances, such as being ill or injured. Most of these men were given a pink slip with orders to move out of their homes immediately.

Then there was the 'portal pay'. The men worked an eight hour shift in the mines, but might spend as much as an hour waiting for the 'man trip' to bring them out. All the mine cars had to be dumped before we were brought out. There were numerous other things as well that we ironed out in the three years I worked there. The last one was at the shaft mine and was covered by another Local 5960 Union.

The first summer I worked in the mines, we had a large statewide meeting in Charleston, W.VA, to hear and meet John L. Lewis and to have a wage meeting with a representative from Union headquarters in Washington, DC.

We got up at 2:00 A.M. on a Saturday morning. A young man, Lord Hicks, was hired to drive us to the meeting. My father, Mr. Kane, Mr. Miller, and I went. It was a long and tiring trip, over bad roads, potholes and such, but it was quite a joy to see the golden dome of the Capitol Building and to have breakfast and lunch in a café on the banks of the Kanawha River in downtown Charleston. It was exciting to watch the pontoon-equipped planes take off on the river, coming under the bridge before soaring up and away. There were even hangers on floats, like barges.

After an hour and a half of business, John L. Lewis gave his speech. It was uplifting and positive. We all listened closely because we knew that here was a man who knew what he was talking about. He had entered the mines as an eleven- or twelve-year-old, as a trapdoor opener and was the son of a Welsh coalminer. A trapdoor opener spends his whole shift opening and closing large wooden doors that span the main tracks in the tunnels where the cars go in to be loaded and are brought out to be dumped.

These doors keep the forced air moving across from one work area to the next and the air is permitted to escape nearer the outside in the main tunnel shaft.

The ventilation in a mine must be kept as near perfect as possible to avoid the buildup of methane gas and coal dust, which can cause a devastating mine explosion. A spark from a pick or hammer would easily set it off.

We were sitting down at the left front of the auditorium. After his speech, Mr. Lewis came down and shook hands with us. The moment I grasped his large hand and felt the firm grip, I looked into his eyes almost hidden under those large bushy eyebrows and was instantly overcome with joy and a deep admiration. Here was a man without a high school diploma or a college degree, yet he had the best vocabulary and spoke so eloquently and gave such an inspiring speech. There was a sort of aura and a great personal magnetism about the man. I knew he was a great and honest man and that we were going to succeed in bettering the working conditions for all coalminers.

"It is a pleasure to meet you at last, sir," I said.

"I am glad to meet you, too," he replied. "And together we have a lot of work to do!" Then others crowded around to shake his hand. It was a very jubilant group going home from Charleston.

The Coal Operators offered John L. Lewis $50,000 a year to be the counselor, which was twice the salary the Union paid him. Over the years, John L. had traveled by car in a convoy of three or four cars. All the overpasses were carefully watched. A lot of Coal Operators wanted him dead. I know that a lot of my boyhood friends said that when they prayed, they prayed first for John L. Lewis's safety, then for their Mom and Dad! I felt the same way. He was the only person risking his life for us, trying to make our lives better.

One day, in a Union meeting, a member said that The Daily Worker newspaper was condemning John L. Lewis in a not very complimentary article. I was asked to answer the man's question. I told him and all the members that we all have the right to express our opinions and so do the people who write articles in newspapers. Of course, I pointed out, they cannot be libelous and such. Even a Bolshevik or a Communist is allowed to have a newspaper in the Untied States. I noted that this individual reacted in an odd way and wondered where he had read the article and if he had the newspaper.

Two weeks later, I received a letter from Max Himler, the editor of The Daily Worker. He informed me that he would have their lawyer proceed with a libel suit against me unless I retracted my statement about his newspaper.

I read the letter to my father. He silently left the room, went in to his bedroom and straight to a large trunk where he kept all our records and important papers. He came back and handed me a stack of worthless stock certificates—$1,200 dollars worth—from a coal company venture he and hundreds of other miners had invested their life savings in. That company built a spur off a main railroad line and a steel railroad bridge leading to the proposed coal company site.

They spent $60,000 and went bankrupt. The signature as president of that company on the stock certificates was Max Himler, the owner and publisher of The Daily Worker!

I contacted a family friend who was a lawyer, Thomas E. Lovas, in War, W.VA, before the next Union meeting. Tommy's parents were also Slovaks from near Bratislava, where my parents came from, and had come here around the same time. Tommy was my sponsor when, at age twelve, I was confirmed and given his name, Thomas, as my middle name. The family owned a large department store in War.

Tommy read the Himler letter and began to rock back and forth in his chair, laughing and having a great time! I became concerned and asked him what was so funny. Tommy replied, "Joe, this is the funniest thing I have ever read! How can a boy 18 years old, going to high school during the day and working in the mines at night, giving all his money to his mother to help the family be sued? Sued for what? The Daily Worker is a Communist newspaper and they are often referred to as Bolshevik Party by their own people!" He looked at me. "Don't worry about it. I'll write Himler a letter as soon as my typist gets here and I am positive that will end the matter."

After all the formal business was taken care of at the next Union meeting, I stood and read Himler's letter to the members of our Union. They were all shocked and threatened to beat up the person who wrote to The Daily Worker if they could identify him. As I read the letter, I kept looking up, directly at the member I suspected was the culprit. He turned red in the face and kept looking at the floor.

Our State Representative, Bill Jenkins, from the American Federation of Labor, was visiting our Local that night and rose to his feet. "Forget it, Joseph," he said. "They don't have a case and they know it!"

Then I told them of my visit to our family's lawyer and what he had said. They all had a good laugh. One miner in the back of the room called out, "Joe! They can't get blood out of a turnip!" That seemed to sum it all up and we all had a good laugh.

On Monday mornings, we started getting snide remarks from the new foreman at the man trip during the summer I worked the day shift. Someone was giving him all the information about what transpired at our Union meetings, which was supposed to be confidential.

I asked a couple of the young men whom I trusted to sit in the back of the room at the next meeting and leave halfway through the meeting. One had a car and I asked him to drive to the foreman's house, which sat on a flat spot on the side of a hill, and park nearby. They did and they saw Carmen Torres, Sr. and his brother, Tony, walk up to the foreman's house and go inside.

These were the only ones who seemed to object to everything we tried to propose to better conditions in the mines. They were always complaining that the Union wasn't a good union.

Our officers got together and discussed the issue and decided to call the two men in on the pretext of giving them positions on the Wage Scale Committee in a couple of weeks.

Meanwhile, our Financial Secretary, George Palus, took a week off to drive to Johnson City, Tennessee to visit and romance a beautiful redhead (I had seen her picture and she was a doll!). The Torres brothers immediately started passing the word around that George had used Union funds for his vacation and romancing. I knew George wouldn't do anything wrong with the money collected for dues because we worked together a whole year studying that radio correspondence course and we knew each other very well.

At the trial, which was a special meeting of key personnel in the Local 5959 Union, the Torres brothers were asked to present their evidence showing that George had misappropriated funds.

They had no evidence but said that they were told that George used the funds. When asked by whom, they could not furnish that information either. We called George in next. He brought his ledgers, bank statements and receipts, all of which were very neatly and accurately kept.

After a brief consultation, the members found George not guilty, of course, and, in fact, commended him on his work. As he walked out of the meeting, George called the Torres brothers "lying S.O.B.'s". It took more than a few of us to restrain the three of them!

We asked the two "Torres brothers" to stay for a few minutes and went into the problem of having our Union business information leaked to the mine foreman after each meeting.

The brothers looked at each other in disbelief and shock, then said, "Who do that kind of dirty backbiting thing?"

President Kane replied, "You did, Carmen and Tony! We had men posted at the foreman's home and they saw you and are ready to testify before the whole membership at our next meeting, a month from now." He looked at the brothers. "I don't think it will be safe for you here after that meeting. I'm sure the whole county will know about this. So please find a place far from here that will be safe for you."

The "Torres Brothers" left, stomping out of the building, cursing and shouting, in a terrible rage. They moved about three weeks later.

George Palus resigned from his position as Financial Secretary and my father was elected to take his place. Dad insisted that he be bonded so there would not be any more problems. The trial of George Palus would come back to haunt him, which I will talk about later.

Our biggest problem after that was to keep the lid on the young members who were always ready to go on strike for the slightest reason, but the older and wiser men prevailed, much to my joy. I remembered the Boone County strike all too well.

The Coal Company officials were, as a rule, meeting our problems and resolving them nicely, but they refused to renew our contract at both Caretta mines. We went on strike.

The Company had been bought by Consolidated Steel, which made these two mines captive mines. Things got ugly in a few days. Everyone carried concealed handguns and there was even a machine gun on the shaft mine bathhouse.

The Company brought in a large group of coal miners ("scabs") from Alabama by train to the Caretta Shaft Mine. Their leaders took one look at the men waiting there for them and realized there was no place for them to stay, they went back to Alabama. That was the silliest, dumbest bluff I had ever seen, one that could have cost a lot of lives.

At one point during the confrontation, a miner bent over to tie his shoe and his .32 automatic revolver discharged, blowing a hole in his hip pocket, taking a little flesh with it. The man did the wildest dance I'd ever seen. It took a few men to restrain him so he could be given medical aid. That little incident almost triggered a battle with the so-called "scabs" from Alabama. I can still remember how dry my mouth was after that! A few days later, the issue was settled, much to our satisfaction.

Consolidated Coal defaulted on its payment to Carter Coal Company and Carter got back both the Caretta and Coalwood mines, the communities, and all the improvements. The story was that Consolidated owed 90 million dollars, spent $5 million in improvements in three years, and had lost $15 million in investments. That was really big money in those days when a loaf of bread or a quart of milk cost eight cents each!

About the time we settled our strike, a new union tried to invade the Untied Mine Workers area. It was the Industrial Workers of the World and was reportedly backed by the Communist Party of the USA. They spent a lot of money trying to secure a foothold in the southern coalfields of West Virginia but failed. Too many of our coal miners knew about it and did not want any part of it. Most were foreigners and felt that they had severed ties with Europe and were more interested in democracy.

Again, the older foreigner's cool heads prevailed. These rumors of company unions flew, but we knew from experience that this was about as safe as the old story of "the fox guarding the henhouse" and about as fair as the weighing of the coal at the scales by Company weigh men!

The coal miners in Twin Branch, owned by the Ford Motor Company, wanted Union representation and a contract, which was refused. They went on strike and after six months, they were quite hungry. We in other mining towns sent them money and supported them as best we could. After negotiations failed, the Twin Branch mines were closed and all the rails, ties, houses, etc., were sold as scrap and the entrance to the mine was sealed. Of course, the families had only a short time to find another place to live and work, but a prosperous town had ceased to exist.

It really didn't make any sense to me to strike when you were earning $5.00 a day for a contract to pay $3.56 a day minimum wages. Even my classmates in Business Math in high school thought the Twin Branch coal miners were short on brains, or at least had been standing behind the door when the brains were passed out!

After we organized our Union, we heard about other towns where the coal miners had hospitalization and death benefits in their contracts, with small monthly payments deducted from their pay each month. The four of us Union leaders traveled to Welch, W.VA, to the closest hospitals to our town, to see what we could find out about negotiating a deal or contract. I had written to all three hospitals for appointments for the same day, at different times.

Our first appointment was at Welch Hospital, which was a really nice place but didn't impress us very much as far as emergency treatment went. Then we went to Grace Hospital, where we found the same thing, and finally to Steven's Clinic Hospital. At Stevens, we were greeted with open arms. We were very impressed with their equipment and the availability of radium for cancer treatment, which neither of the other hospitals had. Even though the radium was kept at the parent hospital in Bluefield, WVA, around 30 miles away, it was always available for Stevens Clinic.

If we signed a contract with Stevens Clinic and our shaft mine and the two Coalwood mines did the same, the Clinic would cancel the $33,000 debt owed by the families from the four mines. All four unions signed and accepted the contract.

The deduction for a family was one dollar a month and $1.50 per month for a single man. Hospital records showed that single men had more hospital care because of accidents, fights, and carelessness than married men and their families.

The Coal Company gladly accepted the contract and deducted the money from our paychecks. It took a lot of pressure off the Company doctors.

In addition, we had another 25 to 50 cents deducted from our pay to cover funeral expenses, which had a limit of $350. This helped a lot, and any funeral home could be used. Up until then, the Company had provided a box and a truck to carry the dearly departed to the Company cemetery, a small fenced-in plot on a hillside just outside of town. The men who did the digging and repairs outside the mine usually dug the graves. A priest or self-ordained preacher would conduct the services, which were short. Markers or tombstones were the responsibility of the family.

When I was about eight years old, I went to the funeral of a Company deputy sheriff, Ben Miller. A bootlegger or someone out for revenge for previous disagreements had shot him.

It was the custom to open the box at the graveside and people would file past and say something nice about the deceased. The problem was, no one liked Ben Miller! In fact, he was one of the most hated people in both towns, arrogant and overbearing, a man with too much power.

As each person walked by the open casket, most of them just shook their heads from side to side but made no comment. Ben's chief, Bud Booth, was at the end of the line and let everyone cut in front of him. Finally, it was his turn. He looked down at Ben Miller and in a low voice, said, "One thing good about ole Ben Miller, when he died, he did a damned good job of it!" I have often wondered if that would be all someone could force himself to say at my funeral.

Of course, the Bible says, "Methuselah lived nine hundred and sixty-nine years and he died." What did he accomplish in all that time? Is that all there is? We all would like to believe that the world has benefited in some measure by each of us being here, before we depart into history.

There was a Polish lady, Mary, who died. She was the wife of Stanley Petroski, who was a coal loader. They had five children. Mary weighed over 300 pounds when she died. The $350 allowed for her burial was insufficient to cover the huge size of the cheapest casket so a special one was made.

Stanley spent his entire savings for that casket and a twin burial plot in Premier, WVA, some fifteen miles from Caretta. Then the priest needed $25 to conduct the services at the graveside, which was only to cover his expenses from the church in Welch and back. If Stanley couldn't pay him, he couldn't do it.

My father and a couple of his friends hired two gravediggers to dig the grave. On the day of the funeral, my father and I picked up Stanley and his children at their home and took them to the funeral in Premier. There were quite a number of friends of different nationalities at the funeral, but no priest or minister. A Hungarian man stood by the grave and read the prayers for burial from a Hungarian Lodge Book. We all knelt then and recited the Lord's Prayer, each softly in his or her own language. Then everyone stood and walked single file by Stanley and his children.

On the drive home, Stanley thanked us, saying, "The funeral was nice. Me no see better one!" I thought it was an unusual funeral but a nice one.

Later, we collected money to help the Petrosky's pay off their debts and Stanley went back to work in the mine. They managed to keep the family together. Two years later, Stanley married the widow of a man killed in the mines. She had two children of her own.

The second summer, the foreman of the mine denied me a fulltime job for about three weeks, because of my Union activities or so a person who was close to the foreman told me. The Union could do little because the Company was not hiring any new people. My father happened to see Mr. Jennings, the general manager of the four mines, at the Caretta mine office on one of his regular visits from the main office in Coalwood.

They shook hands like old friends and stepped out of earshot of the other people around. My father told Mr. Jennings about my problem and Mr. Jennings said he would take care of it. Then he asked about my mother and the family. As he stepped away, he turned and said, "My wife will be over next week to get those peppers, Mr. Diviak. Okay?"

"That is okay." My father replied. I had my job back, but on the night shift and not under that foreman.

Work in the shaft mine was better in regard to the coal. It was a six-foot seam, and I could stand up in it, but it was a very gassy methane mine. At night in the work places, you could hear the gas blowing through the water, making all sorts of crazy sounds. Often it sounded like someone screaming, or someone clearing his throat, or walking closer and closer but never appearing.

Everyone seemed to have a working partner, even the coal loaders, which was nice in case one of us was injured or needed help. My first buddy was a young man, Jim Dillion, who had just been released from jail. He had been charged with capital murder but found not guilty after spending a year in jail waiting for trial and had just been rehired by the mine.

The incident had happened at a roadhouse near English, WVA. Jim and a local coal company sheriff were vying for the favors of Bessie, a "lady of ill repute", as it was called in those days. The deputy came in and saw Jim and Bessie holding hands in a dark corner of the main room of the place. At the time, there were several young local men waiting to take the deputy down a few notches because of some past problems they had with him.

The deputy sheriff went up behind Jim and hit him in the back of the head with a blackjack. Someone pulled the main light switch and everyone was trying to get out the one door at the same time. When the lights came back on, the deputy was dead, with seven stab wounds and the knife still stuck in him. Jim was out cold on the floor. He recovered in about twenty minutes, with the help of a company doctor. Jim was jailed and charged with the murder.

Jim and I worked together for nine months and I watched him closely. Then he decided to leave Caretta and start over in another town. As we shook hands and said goodbye, Jim said, "Joe, you watched me closely but you never once mentioned the trouble I was in. You treated me like a brother. I would never hurt you and you knew it. Take care and be careful, ole buddy."

In between, I worked with different buddies. We drove tunnels through the solid rock faults to pick up the coal seam on the other side, about two or three hundred feet. We ran a 45-pound jackhammer to bore the holes in the rock to blast the rock loose, then loaded it into mine cars to be taken away with other cars filled with rock and slate. They were hauled to the elevator, hauled up top and outside to an aerial tramway of large containers that looked like cable cars in Switzerland. The rock was carried up the mountain to a valley and dumped in a huge pile.

Woodrow Thompson, a large young man from Wise County, VA, became my buddy and we were really good at what we had to do. I learned how to use the "jelly bell" dynamite and caps with the wooden tamping stick, clay dummies, and 100 feet of cable and a detonator. Each night, the section foreman gave me a spare safety lantern that could detect gas. After we had drilled three holes four or five feet deep at an angle downward, I would blast one hole after the other.

For some reason, the smoke and fumes didn't give me a headache like it did most of the people on the 30-man construction crew, so I was selected to do the blasting.

One night, we went to the large concrete powder storage vault outside the shaft and I picked up a case of dynamite (144 sticks) and Woodrow got several boxes of blasting caps. We took the stuff to the cage (elevator) and everyone else stepped back out of the way. I sat the dynamite down and got on the elevator alone and went down the 600 feet to the coal seam level. I got off and sent the elevator back up. Woodrow put the case of dynamite on the elevator and sent it back down to me. I unloaded the dynamite, sent the elevator back up, and Woodrow sent down the blasting caps. He came down on the next trip.

This was a matter of safety in case the elevator cable broke or the elevator got stuck. We carried the dynamite and caps down a side tunnel parallel to the main tunnel, away from all the traffic. We came to an arched overpass on a cross tunnel that was eight feet high. No problem! I helped heave Woodrow up so he could grasp the edge of the flat concrete roof of the cross tunnel. He made it okay and I handed him the caps, then the dynamite, and then he hauled me up. On the other side, I dropped down and Woodrow handed me the blasting caps. I set them aside. As he lay there on his belly, hanging over the edge with the box of dynamite, I raised both hands in position to take the box. His wet hands slipped and I caught the box in the cradle of my arms, against my chest. I stumbled backwards until I landed on my rear end in five inches of the coldest water I had ever felt!

There I sat with my lower body freezing and my head, face and chest covered in cold sweat. I was shaking all over. Woodrow dropped down and ran over to me. He took the dynamite and put it in a dry spot, grabbed my hands and pulled me to a cold standing position. You can burn dynamite and nothing will happen, but a shock will set it off like a blasting cap! The stuff we had contained more nitro than the coal miners used for coal. We were blasting solid rock.

We picked up the stuff and walked a few hundred yards to our workplace, which was warmer than the main tunnel or side tunnels. We never told anyone about our close call that night but we learned a valuable lesson—to be more careful. There were a lot of close calls but it seemed we got used to them.

One time, my boss was sitting on a large rock that I was breaking up with a 16-pound sledgehammer. I heard a faint cracking sound and yelled, "Jump!" He jumped toward me as a bunch of rocks came down, landing on the spot where he had been sitting. You always think accidents happen to other people. My boss said, "Thank you, school boy!"

After running the jackhammer all night, with earplugs and a respirator on your face, the next day in class, all you hear is a roar in your ears. You can see the teacher's lips moving but you can't hear the words. Until you learn to read lips a little, you had no idea what the teacher was saying!

We had a wonderful large console radio at home that we had ordered through the Sears catalog. Once a month, on a Saturday night, I would try to hear the Grand Ole Opry out of Nashville, Tennessee. My mother would come into the room holding her ears, so I would have to turn the radio down to where I could hardly hear the program. The radios then didn't have automatic frequency or volume controls. The sound would burst out or drop so low you couldn't hear it, or the station would disappear for a fraction of a second. But with a better antenna and the cold crisp nights, it was great listening!

Gilbert Brown, our History teacher, had mercy on us boys who showed up for class with black "eye shadows" around our eyes. Even lye soap wouldn't remove the black. Mr. Brown would not call on us too often in class. I was usually well prepared because I read my notes in the mine during our midnight break for food and rest. I had the most problems with Latin. As the kids used to say, "It killed the Romans and it's about to kill us, too!"

Mr. Huff had a "mushy" voice. Even with my normal hearing, it was hard to hear, and it was almost a hopeless battle with him. But being a "foreigner's kid", I learned the vowels as they say them, which helped in Latin. I really studied the book more than the rest and managed to pass the subject. Much to my surprise, it has helped me a lot in spelling and pronouncing words correctly.

If we'd had a counselor in high school that we could have gone to ask questions about college and costs, it would have been a huge advantage for us to make a correct career choice. All I knew was that to get into the University of West Virginia, I needed four credits each in English, math, history, and science. I had that. I also needed one year in a foreign language. I had that, too. I also had four electives, which also would have helped. But no one seemed to know what the cost would be. I wrote to the University and received a lot of information, but nothing about the cost of living or rooming off-campus, etc.

I wanted to take Radio Engineering, but the closest thing they had was Electrical engineering, which dealt with all types of motors, dynamotors, galvanometers and such, all ancient stuff, but nothing on radios. One day I was reading the Cincinnati Post, that one of our neighbors had given me. There was an ad about the Army Air Corps Technical School in Rantoul, Illinois, known as Chanute Field. I wrote them a letter and mailed it before I graduated from Big Creek High School in May 1935.

There were 83 of us in the senior class of 1935. I ranked in the middle of the class. The graduation exercises were held in the evening in the auditorium. My class sat down in front, to the left, and we had to go up the steps to the right and then cross the stage to receive our diplomas. We were all dressed in our caps and gowns.

Our principal did a great job of getting the program under way. After the invocation and the Pledge of Allegiance, we had the speaker, a prosperous businessman from Charleston, then the valedictorian and the salutatorian, and finally, the parade of graduates across the stage.

I looked out of the corner of my left eye at my mother as I went to get my diploma. She and my oldest sister, Theresa, were wiping tears of joy from their eyes. That sight made all the hard, tough years I spent getting here worth it. It was a mountain and I was the first one at my end of town to climb it.

After tearful farewells with our classmates and teachers, I drove home with Mother and Theresa. My father greeted me with my first bear hug ever from him and a handshake and the most wonderful words: "You make it yourself, Yosef!"

Monday evening I was back in the mines. I needed more money to go to college. I had a few hundred saved and was earning more money as a coal loader. I still gave my mother money to help with my younger brothers and sisters.

While waiting to hear from the Army Air Corps Radio School, I really didn't want to take too many more chances in the mine, but I had no other choice.

One night, Woodrow and I were loading a car with coal, our third, when we heard a crack in the next tunnel. I ran around the corner through a breakthrough tunnel to see what had happened. I saw one man standing rigid as a statue. All I could see of the other man was his feet and legs protruding from beneath a large slab of slate that had fallen from the ceiling.

I grabbed a crowbar and lifted the slate up while Woodrow pulled him out. His name was James Harris and he was also working to go to college. He was one big bloody mess. The section foreman came quickly and then a couple men ran up with a stretcher. They loaded James on the stretcher and put him on a coal car like the one we had been loading. The car was hooked to a six-ton locomotive that rushed him to the elevator, about six miles away.

A doctor met him at the top and was able to stop a lot of the bleeding and then James was rushed off to the hospital in Welch. James survived but he had horrible blue (coal) cuts in his face, some broken ribs, etc. He had to have plastic surgery in Charleston to remove the coal-blackened wounds.

We kept right on working after they took James away. Later, the foreman and section foreman came by and asked a few questions about the accident. They picked up the 6-foot long steel crowbar, which was now bent, and tried to lift the slab. They couldn't. The foreman looked at me in disbelief when James's partner and Woodrow told him what I had done. Later on that shift, I tried it again but couldn't lift it as far as I had before.

James came by the bathhouse two months later and we had a nice chat.

CHAPTER SEVEN

Going to College

In August, a representative from the National Business College in Roanoke, Virginia came by the house and told me that some of my buddies had signed up to go to college and were taking banking and finance courses and salesmanship courses. I signed up after seeing the deposit receipts of my close friends, Fred Vogelsong and Donald Ceunelo, both from Coalwood. I was surprised at myself that I did this.

I continued to work in the mine until the enrollment date in September. When I graduated, I had sent a commencement announcement card to Ruthie Horton in Woodlawn. I had seen her several times at her sister's home near the Company store in Caretta and had once picked up her and two of her girlfriends in our new Studebaker near the shaft mine in a rainstorm. I had driven them to Ruthie's sister's house. I had spoken to Ruthie a few times as I passed their house on my way to work or school.

Ruthie responded to my commencement announcement with a beautiful graduation card and that began a loving correspondence when I attended Business College in Roanoke. I also had received money and cards from Mrs. Jennings, the general foreman's wife in Coalwood, and a few others, which I really appreciated.

I took a leave of absence a week before college enrollment and just relaxed, enjoying the sun and trying to get rid of that pale miners look and get the coal dust out of my throat. I had lived for nineteen years in the coalfields and it was time to further my education. My two sisters and two brothers were doing fine in school and were all looking toward graduating from high school like their "big brother" had done.

My older sisters didn't have a chance to even get to junior high school. They went to work early. Teresa and Mary worked as domestics in boarding houses; Anna at 17 and Helen at 15 went to Perth Amboy, New Jersey where we had some friends that helped them find jobs. Anna worked in a cigar factory and Helen in a large department store.

Teresa, Mary, and Helen were all married in Caretta, with nice church weddings and a celebration at home with orchestras and all. Teresa married a Yugoslavian, Mary a Polish man, and Helen a German. So everything had worked out wonderfully at last and now it was time for me to do what my mother and family told me, in their Slovakian language: "Do your best, Yosef, and we will be proud of you!"

I caught the train for Roanoke, Virginia in Welch. It took about four hours to get there after all the stops and waiting for other trains going in the opposite direction to pass. It was a short walk from the station to the YMCA, where there were dorms on the second and third floors.

I was assigned to a room with a young man from Rocky Mount, Virginia. His name was Delma Young (evidently, his mother had wanted a girl and he was an only child!). We became great pals and had a good time there. We ate three meals a day down the street, up on a second floor boarding house. The food was delicious and no one ever complained.

The first morning in college, we were assigned to our classes, which started the next day. Courses included bookkeeping, banking and finance, commercial law and English, typing, salesmanship, and most of all, a heavy emphasis on one hour of penmanship. We made pages and pages of each letter in the alphabet and they had to be perfect or you did them over and over again. I think it should be a requirement for all medical schools to have that course! Ours was the Zainer-Blosser Method. Other schools taught the Palmer Course.

We all worked hard each night on our homework and about 11:00 P.M., six or seven of us would go down the back stairs to the indoor swimming pool. We swam in the buff, but didn't dive or talk or make any noise at all. We just swam in a circle around the pool, and then we would dry off, go back upstairs to a hot shower and bed.

On Sundays, we played basketball on the roof, above the sixth floor. There was a basketball court enclosed by wire. We needed the exercise to release the tension of so much schoolwork and homework.

On Saturday afternoons in the fall, we would go to a college football game now and then. The Christmas holidays at home were wonderful and everyone was doing fine. They were all proud of brother Joe. Then it was back to college after the New Year and the college routine.

I had been corresponding (love letters!) with Ruthie. She was at home in Woodlawn, Virginia, about ninety miles southwest of Roanoke. She invited me to come for a visit and would pick me up at Galax, Virginia in her stepfather's Model "A" Ford and drive me to their home in Woodlawn. The plans were set for the following Saturday. My train would arrive in Galax at 11:00 A.M. .

I'll never forget the coach I was riding in. It had a large iron coal-burning stove (we kids called them "pot-bellied stoves") that stood in a sandbox in the center of the train car, with the stovepipe going up through the ceiling. It was cold and snowing but the train made it through to Galax. What a sweet greeting with a nice kiss and a hug from Ruthie. She drove us to her home where I met her parents and a sister and a brother.

We had the living room to ourselves that afternoon and evening. The next morning, we took a walk outside but quickly came back inside. There was snow on the ground and it was very cold, with the wind blowing. We sat in the living room and talked and just enjoyed each other's company.

We ate with her family—the meals were delicious, the best home cooking there on the farm.

On Sunday, Ruthie drove me back to Galax Train Station. We hugged and kissed and said goodbye. When the train pulled into Roanoke, I wondered why there were so many people standing on the platform, shouting and waving. I found out quickly enough when I stepped from the train.

All my classmates, boys and girls, were welcoming me back and shouting that they wanted to "see the bride"! My roommate, Delma, had let my secret out. I took a lot of kidding but knew it was all in fun.

Ruthie and I kept writing to each other. She had returned to high school and was trying to be a valedictorian, while doing four years in three, which she did. Ruthie had dropped out of school five years before to go to work.

I was asked to join a fraternity since my grade point average was a 3.5 out of a possible 4.0, but after hearing about all the silly hazing and other degrading things from the upper classmen, I turned it down. I didn't want someone's initials imprinted on my rear end from a monogrammed personal paddle! I sure didn't need that! The boils on my back, from injuries in the mines, had all healed. The sores from the battery acid leaking from the headlamp batteries in my left rear pocket area were still tender though.

The fraternity was shocked when I turned them down and kept after me to join. But I said, "No. I am not interested. I ceased being a boy when I went into the coal mines and did a man's work over three years ago."

In May, college was over. I had completed all the courses except bookkeeping and banking, which would take me another year to secure my "AA" or Associate of Arts degree. I returned to the Caretta Shaft Mine and my buddy, Woodrow Thompson.

It was tough getting back in shape, but within two weeks, I was back to 169 pounds or, as they said in those days, my "fighting weight." I needed more money for college so I loaded coal and did well at it.

From time to time, there would be a "mountain break". In a large section of mined-out area, the tons and tons of rock from the ceiling would snap all the timbers, causing a shockwave that scattered small lumps of coal and dust sideways all over us. We were working the "pillar line" at the time and we would laugh, shake ourselves off and keep on working.

Woodrow kept telling me about his wonderful Juanita and I told him about Ruthie. It helped us to work harder to make our dreams come true. It gave us hope and something to look forward to and time seemed to pass more quickly. I found out some fifty years later that Woodrow and his Juanita were happily married for five or six years before Woodrow was killed instantly in a "mountain break", where he and I had been working together when I left the mine.

In August, I received a package of papers from The Air Corps Technical School in Illinois. It had all the information I needed and instructions that, if I was still interested, I should take the enclosed medical forms to the Company doctor and the dentist, then submit them to the school. It would save me a trip if I were not physically fit.

I immediately took the forms to the doctor, who checked me over and said I was perfectly fit. He filled out the form. Then I went to the dentist, who also examined me and filled out his form. I mailed both forms back to the school as quickly as I could.

In the booklet that came with the package, it explained that there would be two days of written exams and one day of physicals. I was to bring all of my school records, etc. It also stated that, after a year, if I was not satisfied, I could "buy out" of the Army Air Corps and would be given an Honorable Discharge.

I made sure I had the money available, in the bank, just in case I wanted out. Meanwhile, I went to War, WVA—a town of about 1500 people back then—and spoke to Mr. John Martin, manager of the Bank of War. It was quite a name! He had a cashier and bookkeeper that earned about $75 a month and jobs were scarce.

I gave up the idea of going back to Business College in Roanoke, although one of the best things I had done was to go there. I found out later that the year I had spent there and the things I had learned were invaluable when I began doing business with insurance companies and others. My learning kept me from entering into business deals without a contract and made me aware of scams and crooked deals.

A few days later, I received a letter to be at Chanute Field at 9:00 A.M. on October 21, 1936. I had thirty days to wait, so I quit the mines and sold all my coal loader's tools. Mr. Irving, my boss, offered to lock them up for me in a secure place in the mines, but I told him I was never coming back to work in the mines again!

I was ready, mentally, physically, and emotionally, to move on with my life. He thanked me for the hours I'd spent on my midnight breaks helping him design an overshot water wheel to generate electricity on his large farm in Asheville, NC. I'd had courses in physics and science, which helped us both a lot.

My boss needed electricity for his farm. We came up with the idea of using two large old wagon wheels, welded to a long shaft, spaced about two feet apart on one end. We put a large belt wheel on the other end. The wagon wheels were planked on the outside with aspen lumber to make a ten-foot diameter water wheel, with troughs all along the outside to catch the water.

He had a generator and lots of wet cell batteries, which he kept charged all the time with the water wheel drive. By switching the belt off the generator and onto a grinding wheel, he also had a mill to grind cornmeal and flour. It worked out nicely.

We shook hands and gave each other a "bear hug". I said goodbye to him and my other buddies at the mines. It was a tearful farewell.

I used the time before I had to leave to clear my lungs and heal my scars. I also brushed up on math, history, and other subjects that had been mentioned in the Army Air Corps booklet.

My brothers were really worried about my leaving but they were also glad that I was getting out, even though they knew they would inherit all the chores I had done, just like they had when I went off to college.

CHAPTER EIGHT

In the Army Air Corps, Chanute Field, Illinois

It was a sad farewell with my parents, but they were very proud. My brother-in-law Fabian drove me to Welch so that I could catch the train to Rantoul, Illinois.

When the train arrived in the small town of 1200 to 1500 people, I went to the hotel and secured a room. The next morning, I walked the mile to the headquarters where a grey-haired Staff Sergeant greeted me. He asked if I was staying at the hotel and I replied that I was, which was fine with him.

At 9:00 A.M., there were about forty of us sitting in the large classroom. We were given a test on high school subjects, etc. We had lunch and went back to the classroom until around 4:00 P.M.

We were all back again the next morning. This time we were separated as to what we were interested in. All the prospective radiomen were in one room, the aircraft mechanics, welders, parachute riggers, sheet metal workers, etc., were in other classrooms.

We finished at 4:00 P.M. that day and had to be back again the following morning by 8:00. I had passed "with flying colors", as the old Sarge said. All I needed to do now was to pass the flying physical that day.

I passed, except that I had an infected big toe and an ingrown toenail. A suitcase had landed on it in Cincinatti, Ohio. I'd never had any problems like that before. Captain Cameron, the flight surgeon, said, "No problem." After I signed a waiver, he injected the toe with novocaine and removed the toenail.

I earned my keep in the infirmary, while I was recovering. I sat in a chair with my foot elevated on a stool with a cushion and peeled potatoes by the sack full! The "crew" kept me well supplied with potatoes and emptied the peelings and rejects every hour or so. One kind lady nurse would bring me a large mug of what looked like coffee, but was really black beer to help "kill the pain" and throbbing in my toe. They sure don't have nurses like that any more! She was a real lady!

While in the infirmary, I met Jim Muri, a fellow applicant for the Aircraft Welding and Sheet Metal course. He was put in the infirmary because they found too much sugar in his urine. He had eaten candy bars for two days while he traveled here from Miles City, Montana.

When we were released from the infirmary, we were sent to a WWI-type barracks. There were enough of us to form a "boot camp" and begin our training. We only had to be sworn in first. I went out to the main gate and back twice. The third time I started out, it suddenly occurred to me that I had to go through with it. I was in better shape than most of the other thirty-nine men, other than Jim, who was also in excellent physical shape.

The Air Base Commander, a Colonel who was really nice and friendly, swore us in. He looked at me as he shook hands and noticed my pale white skin. "Ex-coalminer?" He asked.

"Yes, sir." I replied.

The Colonel nodded. "You'll make it."

The Army Air Corps, in 1936, was still part of the Signal Corps and regular Army, but it was becoming "too big a baby for the Signal Corps to nurse", much to the consternation of the Army and Navy. Plans were in the works to separate the two. Brigadier General Mitchell had personally flown a Keystone bomber (bi-plane) out over the Atlantic and had found an old surplus battleship anchored and hidden there during some war games. He dropped a 500-pound bomb down the smokestack and sunk the thing, thereby dispelling any idea that a bomb couldn't sink a battleship, as the Navy had been saying for years. The Navy kept insisting that we didn't need a long-range Air Force, that they were the first line of defense! That, too, was more propaganda than fact, as was proven five years later at Pearl Harbor when the entire Pacific fleet was bottled up in the harbor, anchored in a place so small that I wouldn't even want to have speedboat races there with outboard motors! What was pitiful was that there was only one access, and a narrow one at that. One ship, if it was scuttled, could block all ships from getting out of the harbor. And, of course, it was a Sunday morning—most were recovering from a night of overindulgence!

General Mitchell was eventually court-martialed for disobeying an order to keep quiet about the new Air Force. The Navy had some powerful friends in Congress, so the nasty job was done. His rank was reduced to Colonel, although in later years, he was exonerated.

We did our two months in boot camp, wearing wrapped leggings over the legs of pants that resembled riding britches. It was a tricky job to secure the wrappings so they wouldn't unravel during the marching and drilling. If we wrapped them too tightly, our legs would become hot and swollen and then numb. We also wore Stetson-type campaign hats. But we did our best. Later, we were able to wear olive-drab slacks and khaki's and billed caps. This was much more functional and made us a better-looking group of men in the Air Corps!

We used the Army .45 caliber automatic on the firing range. Some of the ammunition was so old that the bullets hit the targets sideways or knocked holes in the glue buckets (used for patching holes in the targets) under the targets. After the first shot, I was deaf for hours! I soon learned to take a wad of toilet tissue and ball it up, stuffing it in my ears to deaden the sound. Being a West Virginia "hillbilly" who had learned to shoot in the woods, I soon made sharpshooter and expert once we got new ammunition!

Our Drill Instructor, Bill Wagner, was an ex-Marine Staff Sergeant. He was a fair and honest man, but a tough one. We all respected him and did our very best. He in turn took care of us and worked patiently with us, after the day's work was finished, so that all of us would pass the final parade and inspection by the Commander.

Outside the Air Base's main gate in Rantoul was a night club/restaurant called The Silver Buffet, appropriately nicknamed "The Bloody Bucket" because of all the fights there on Saturday nights, holidays, and a lot of other evenings, too!

We were near graduation from boot camp, so a group of us went there to have a couple of beers and some food. There was a tough-looking Private there, who had nine years of service in and was constantly in trouble, even though he was an expert in his job as an engine mechanic. We called him "Blackie". He jumped two of our fellow recruits outside and beat them up. Sergeant Wagner sent the rest of us back to our barracks while he stayed behind.

About twenty minutes later, he joined us but didn't say anything. On Monday morning, one of the other recruits went to the dispensary for a cut finger. When he came back, he told us he had seen a nurse picking pieces of cinders out of Blackie's face and told us that he also had two black eyes. We knew then what had happened that night outside the "Bloody Bucket"! We never had any more trouble with Blackie after that—he must have gotten the message!

We had our parade and were presented to the Commander in review and passed. Sgt. Bill Wagner had earned another stripe, promoted to Technical Sergeant, and he was happy and so were we. We had a great party later!

We moved into the barracks for students and were in the Headquarters Squadron. Jim Muri and I had top bunks beside each other. Henry Swartz, Jim's buddy from Montana, had the bottom bunk under Jim. Vernon Milligan, from Altoona, PA, had the bunk under me. We all had work detail every Saturday.

Four tired student buddies off of Saturday work details, Chanute Field, Illinois
Vernon C. Milligan, Altoona, Pennsylvania
Henry Swartz, Miles City, Montana
James P. Muri, Miles City, Montana
Joseph Diviak, Caretta, West Virginia
(1937)

There were seventy bunks in each of the ends of the long barracks, with the latrines and showers in the middle. Also on each end was a private room for noncommissioned officer we called the "Barracks Chief". There was a row of bunks going down each side of the long aisle going through the barracks. Long steam pipes that ran the length of the building under the windows heated the barracks. Steam was fed through insulated underground pipes from large coal-fired boilers housed in a large plant next to the railroad tracks, where the large coal cars would unload their coal.

Each soldier had a footlocker. The top bunk man's locker was at the foot of the bunks, on a small special stand. The bottom bunk man's locker was between the bunks under the window. Long four-foot wide mats went down the center aisle in both wings of the barracks.

It was very comfortable and we soon learned to settle down and do as we were told. Supper was from five to six o'clock. We stood "retreat" to the bugle and then marched to the huge mess hall. There were eight men to a table and the food was served "family-style". The man who took the last food off the plate had to go get more. We were allowed to talk, but noise was kept to a minimum because you could get KP on weekends if you misbehaved.

The food was really good and plentiful. After supper, we went back to our barracks and got things ready for the morning inspection. We had a lot of homework to do, which we did lying on our bunks or sitting on the footlockers.

A public address system blared out orders from the orderly room for all formations, including all the bugle calls. After breakfast, it was clean up and get ready for inspection. Then we marched off to school, with small groups dropping off to go to their classes on the way.

The radio course was not only on theory, it included hands-on repair of actual airborne-type radio equipment. Later, we worked in actual Army Air Corps planes based there for that purpose.

Afternoons were mostly practicing and sending or receiving code that started at five words a minute and increased each week, to ten or twelve, then to eighteen words a minute. We had typing on a Western Union keyboard typewriter. All the letters were capitals. No one could print faster than fifteen words a minute and be legible, so we learned to type 30 words a minute.

I was very thankful that I had taken that course with George Palus while we were in school. After the first three exams on theory, radio circuits, and diagrams, the instructor asked me if I had been to a radio school. I did well.

At four o'clock, we were back in formation, then back to the barracks to do our chores of shining shoes and belt buckles, checking our mail, picking up our laundry, etc. At five o'clock, retreat was played while the flag was lowered, then on to the mess hall. As far as I was concerned, I thought the food was well prepared and nicely presented, but we had our complainers, called "bitchers", who complained about everything. But I did notice that they were first in line for mess and the last to leave the mess hall! I figure they were born that way and just never got over it. They led a miserable life doing extra duty evenings and on weekends before they learned to keep their thoughts to themselves! They caused a lot of problems for the whole group and we had to really get on them to "shape up"!

Normally, we worked on Saturday mornings, policing the area and a work detail moving chairs, beds, and such from one building to another, or on guard duty. During June, July, and August, the school shut down so the instructors could take their furloughs and we could paint the hangers, barracks, orderly rooms, and other buildings. We were also allowed a ten-day furlough.

I boarded a bus in Rantoul at four o'clock one morning in August, bound for West Virginia. The bus was crowded and it was a very hot and dirty ride all that day and into the late evening before we finally reached Welch. From there, I took a cab to my home in Caretta. Mother, Dad and my two sisters and two brothers were so glad to see me! They were full of questions and had a lot of things to show me. There had been a lot of changes since I had left. Even my radio shack had been torn down and all my "radio junk" had disappeared. The coal company didn't want any buildings other than homes and outhouses in the hollow. All other buildings had to be out of town, away from the homes. The coal company had re-fenced all the homes, the old fences torn down and hauled away. Everything had to be uniform. It did look a lot nicer.

My father handed me the keys to his Olds and said, "I know you want to visit old buddies and friends." My brother, John, and I drove to War and visited a few friends. On the way back, we stopped at Ruthie's brother's house near the company store. I asked about Ruthie and he said that she was back in high school in Woodlawn and doing great. She had stopped corresponding with me just before I joined the Army Air Corps.

One Saturday evening, a group of young people were having a wiener roast up Saw Mill Hollow, above East Street, and my sister Julia, who was two years younger than me, and I were invited. It turned out to be a real nice evening. The fire felt good as we talked and ate hot dogs and just had a wonderful time.

During the evening, a beautiful young lady and I began talking. She was very interesting to talk to and we agreed to write to each other. We exchanged addresses. Her name was Matilda Morgan. When the party ended, we said goodbye.

Sunday afternoon, my old buddy, Fred Vogelsong, and I drove three miles down the hollow to a theater in Yukon to see a movie. It was very hot that day and soon my nice khaki uniform began to show the sweat, There were no fans or air conditioning in the building, and of course, no windows, so we left the movie early and went to visit some friends in English and Bartley, then went home. I saw my old boss, Bill Irving, and a few others at the shaft mine office and the company store. The days passed so quickly for me.

My brother-in-law, Steve, my sister Teresa's husband, drove me to Welch to catch the train back to Rantoul.

Eight of us radiomen in my class were still unassigned. The rest of the class was men from other air bases who had been sent here to school by their Commanders after competitive exams. The Army Air Corps had eight positions available. Eight small slips of paper were dropped in a hat. On them were written the names of the assignments. Two were Bowling Field, Washington, DC, two from Westover, Massachusetts, two for Fort Sam Houston in Texas, and two were instructors here at Chanute Field. They were expanding the school there.

I waited until last and drew my slip from the hat: Fort Sam Houston in San Antonio, Texas, which was the headquarters for detached air bases in Texas, Oklahoma, New Mexico and Arizona. Ten or twelve men worked at each base. Meanwhile, I would work as a Teletype operator here at Chanute Field while I waited for my reassignment orders to come down, about a month.

In the same area, there was a room where parachutes were issued. A beautifully lettered sign hung on the wall: If the chute does not work, bring it back for replacement. One day, a rough-looking Sergeant came in with a badly torn and shredded parachute and wanted a replacement. It was really funny to watch the expressions on the faces of the two airmen who worked there!

We found out later that the Sergeant had been flying in a two-seater plane. The flight started fine, but at 1000 feet, the engine quit and he was told to jump. But they were too low to jump and the pilot happened to see a wheat field and landed the plane safely. However, the Sergeant's parachute was caught by a strong wind and he was pulled into a barbed wire fence. They took the sign down the next day!

The social hours we had during school were still maintained. There was a diner about a mile from the base, on a side street, that featured a 26-ounce "fishbowl" full of beer for a dime. It was called "Griesedycke Beer Company". One of those "fishbowls", with a hamburger steak that covered a long platter and came with mashed potatoes, gravy, fresh green beans or peas, and oven-fresh rolls, all for 75 cents, was wonderful! The movies, both on base and in town, were great, too.

The Post Exchange had everything we needed, including personal items, at about a third of the price in town. We received $21 a month. $3.50 went for laundry and twenty-five cents went for an "Old Soldier's Home", wherever that was! A pack of cigarettes was fifteen cents. So we all managed to save four or five dollars a month.

CHAPTER NINE

Change of Station to Fort Sam Houston, Texas

Finally, our orders came through. We were to report to the Army Air Corps Air Offices in the Quadrangle at Fort Sam Houston.

We boarded our plane, a Dutch-made C-7, five-passenger plane, where the pilot flew from an open cockpit behind the passenger cabin. It was a single-engine monoplane with fixed landing gear and averaged about 85 miles and hour, ground speed.

The day we left, it was sunny, but there was snow and ice everywhere with temperatures 10° below zero! We had to stop three times for fuel and to thaw the pilot out!

Our Commander, Major Earl G. Harper, came to see us off and shook hands with each of us. When he came to me, he said, "The coalminer from West Virginia and you're the one who shined my saber after had let it go for such a long time! Son, you did a good job! Good luck and I'll see you again someday."

I thanked him and wished him good luck, thanking him for the school there. He nodded and we saluted. I picked up my bags and crawled aboard the C-7.

We stopped at Scott Field in Missouri, to refuel. When we landed at Love Field in Dallas, a truck picked us up and took us to some large tents for the night. We cleaned up and at 7:00 P.M., a station wagon (carryall) picked us up and took us into town to a restaurant near a Theater of Performing Arts. We had a great dinner. Little did I know that the North Texas University State Choir was performing in the theater that night and that my future wife was in that choir that night!

We spent a cold night in those tents even though they were floored and sealed. It was 35° with a cold North wind blowing. The next morning, after a quick breakfast at a nearby restaurant, we were again on our way to San Antonio. We landed at Kelly Field, at the edge of San Antonio, late that afternoon. An open truck picked us up. It was 70°! We shed all the clothing we could but our "long handled underwear", made of wool, was almost too much as we rode through town on our way to the Fort Sam Quadrangle.

We reported in and were quartered with the 7th Signal Company, outside the Quadrangle. The Quadrangle was a fort with high cut-stone walls. A pair of hugs doors made of timbers with large hinges covered the arched opening, and there were holes in the walls for rifles.

Inside the Fort, which was the size of three or four football fields, were trees, flowers, and shrubs with squirrels, peacocks, and other animals. It was a beautiful place. Inside the walls were a lot of offices for the Fort Sam officers and the Southwest Air Corps Headquarters, with windows and doors facing inside the Fort. There was also a large, four story clock tower with four faces on it. The only entrance to the Fort was through the huge doors. They were kept open and a black ornamental iron fence with a gate stretched across to keep in the deer and peacocks.

I became friends with a sergeant who was a radio ham operator. We built a rotating long-beam antenna out of cane poles to work on the ten-meter band. It worked so well that we could talk to hams in Ireland, England, and all over the USA at night! Sgt. James was a great operator who could send and receive 45 words a minute. He worked out at Dodd Field, which was nearby, at an Intercept Station.

While I was waiting for my security clearance to be able to go to work at the Tucson Army Air Corps Airdrome radio station that handled thirty-three government agency messages and served as a relay station, I often went to Dodd with Sgt. James.

We would sit in a room full of listening stations with a lot of receivers. The Sergeant In Charge would tell us to copy a certain frequency at a certain time. They were generally in five-letter codes sent out by other countries to their ships and island possessions. At the end of the day, all the large envelopes holding the messages for the day were put into a large pouch and sealed. An Army Air Corps pilot would then carry the pouch out to a plane on Dodd Field and fly out to deliver it to The Chief Signal Officer's office in Washington, DC. There the messages would be deciphered by a group of experts who had the equipment to do it.

I will never forget the Thanksgiving dinner at the 2nd and 7th Signal Companies mess hall. It was fantastic! There was a bottle of beer and a White Owl cigar by every plate. Dinner consisted of roast turkey with corn bread, oyster and chestnut dressing, giblet gravy, cranberry sauce, baked ham, sweet potatoes with a marshmallow topping, salad, and pickles. Dessert was pecan pie topped with ice cream.

After dinner (which was at noon), we dragged ourselves back to the barracks for a nap!

CHAPTER TEN

Transfer to the Airdrome in Tucson, Arizona

In January, a staff car picked up our commanding officer, Major Pursley, also our pilot, and me and took us to Brooks Field, which was also near San Antonio. We got into a 0-43 Observation Plane and took off. The plane had a sliding canopy over the two-seat cockpit. We could talk to each other through the intercom headphones and mikes.

As we took off, I could see the Shooting Star insignia on the sides of the planes on the flight line. This was the place they had made the movie "Wings" in 1928, which had been a big hit! The movie projected the image that these were fighter or pursuit aircraft, when in reality they were observation aircraft that were armed and directed artillery fire and took pictures of the enemy lines.

We followed the railroad and the highway at about a 2000-foot altitude so that if we had a problem, we could land near the tracks and be found. Otherwise, everything out there looks alike and no one would ever find you alive!

We landed at Dryden Airdrome and the major and Sergeant went over some government property that was to be condemned due to "fair wear and tear". The Sergeant had a five or six-year-old son who's vocabulary included a lot of foul language. His dad tripped over a coiled garden hose and the son said, "Dad is a clumsy old bastard!" It surprised me but I realized that, after all, there were six or seven soldiers living in the house next door to the Sergeant's. The kid would probably be fluent in explicit language by the time he was ten!

We took off right after breakfast the next morning and made it to El Paso. Again, I felt more comfortable with the railroad and, once in a while, a blacktopped road running beside it. The view and scenery was beautiful. It's over 500 miles from San Antonio to El Paso.

We stopped there over night. The Major had more bookkeeping to do there at Biggs Field at the edge of Fort Bless. I met my old Chanute buddies Lewis Howell from Monroe, LA and Herman Schultz from Sioux City, Iowa and we had dinner together in El Paso.

The next morning, we flew to Lourdsburg, New Mexico and then, after more bookkeeping there, to the Tucson Airdrome. Home at last.

When I climbed out of the plane, I was amazed at the beauty of the Santa Rosa Mountains. They seemed to be only a few miles away from the airfield but I learned later that it was at least sixteen miles to the base of the mountains!

Master Sergeant Simpson, the airport manager and Non-Commissioned Officer in Charge of the detachment of three weathermen, five mechanics, and six radiomen, greeted me.

We had our own hanger, radio and weather station building, and our own barracks, which was actually a three-room house next to Sgt. Simpson's home. The airline terminal was across the street.

After I settled in for a few days, the first thing I did was go to the five-and-dime store in Tucson and buy some plates, mugs, flatware, etc., for our kitchen. Knute Shaffer, an old man well over seventy, always had a bunk in the room at the back of our house. Knute was a prospector and a great cook, especially with sourdough biscuits and flapjacks and stews! In the evenings, he would tell us stories of his experiences as a prospector looking for gold, gypsum, and silver, with his burro, in the mountains of Arizona. He had made a couple of strikes but gambled them away, he said.

We bachelors bought the groceries and Knute did the cooking and took care of the lawns around our house and the Sergeant-in-Charge's house.

One night, while Knute and I were enjoying a couple of whiskeys over chipped ice from the ice box in the kitchen, with a splash of tap water in them, I asked Knute if it was true that soldiers in the 1880 to 1900 years really raped Indian women.

Knute took a couple of puffs on his pipe then replied, "I heard of only one case and it was told to me by my father, who was also a prospector. A couple of Calvary men came up on an Indian maiden bathing in a stream, in a well-hidden pool. One of the men quickly dismounted and grabbed her. In spite of her yelling, he began to rape her. The other man rode quickly back to camp.

"All during that afternoon, and during the night, too, the Calvary patrol, too scared to go look for the man, heard a man's screams and moans. The next morning, all was quiet. The Calvary patrol finally rode out and found the man less than a half a mile away. He was dead. He had been staked out, spread eagle, naked and face up, his wrists and ankles tied with rawhide strips to stakes driven in the ground. The Indians had placed him so his crotch was lined up over a large anthill. Then they poured honey over his private parts and rode away, leaving him to die a horrible death. The Cavalrymen dug a shallow grave and buried him. They reported him missing on patrol and presumed dead.

"As usual, someone talked and soon the word got out as to what the Indians would do if any of their women were raped. I never heard of another rape of an Indian woman in my entire life as a prospector."

Knute's story was a shock, to say the least. He told us many, many stories, too numerous to tell here. He in turn enjoyed my coal mining tales about West Virginia.

American Airlines terminal building housed the Department of Commerce offices (now the FAA) and radio stations. There was a restaurant in there, too. It was a nice arrangement. I went to work the next morning as a junior operator. My job was to assist the senior operator by taking voice messages from thirty government agencies in Arizona and New Mexico. I typed them and handed them to the senior and he sent them. The replies were handed to me and I called the various agencies to give them the information. Later, copies of the radiograms were sent to them by mail.

We were on three radio networks: WVM, which was the War Department Network by all the government agencies; WYT, which was the Army Air Corps network for all arrival and departure messages, along with the Navy and Coast Guard planes sent to the last airport they departed and the next airport where they were enroute to land. We never "lost" a plane. The third station was WUM, to link us to Huachuca, Arizona on the Mexican border.

There were eight all-band receivers in a rack at the sending position with a switching arrangement from one receiver to another, as they were needed. They were all on at the same time and you soon learned which stations were calling you in Morse code.

Each radio operator had a two-letter identification letter that was used for every sent or received radiogram. Mine was "DV" because there already was a "JD" in four-state network of about thirty operators. It wasn't long before I lost my name of Joe and became "De-Ve" in our group as well as in our network.

The volume of traffic was a disadvantage to a junior operator. It was overwhelming most of the time and we needed the fastest and best operator on each "trick" to handle the huge daily volume. We said "good morning" on our circuits at 7:00 A.M. and stayed open all day, until 7:00 P.M., when we said "goodnight" (GN).

It didn't take me long to get acquainted with the American Airline radio operator, Charlie Horton, and his helper, Tom Jones, who also did the tickets. There was also Frank Tucker, the chief mechanic and his helper, Bill Giles.

The Department of Commerce (FAA now) had Paul Boatman and Dan Stokes. They had radio transmitters and receivers and kept the five large radio beam towers and transmitters working for instrument flying.

I managed to help all of them from time to time, when I wasn't on trick at the radio shack, and on Saturdays and Sundays. Fran and Earl Johnson ran the restaurant concession in the beautiful adobe and stucco single story building, which had a long tiled porch complete with arches and red Spanish tiled rooms. It was quite an impressive building.

Frank Tucker said to me, "Joe, I should be paying you for all the work you help me with to keep this American Airlines terminal operating, especially on the freight line."

One day, I received a copy of a letter from the Judge Advocate General's office in Washington, DC, stating that I could accept pay for services rendered as long as it was not during the normal military duty hours. Our Non-com Sergeant had sent a letter, along with a letter from Mr. Tucker, asking their permission. I went to work.

Every evening, I took a truck with a large air conditioning unit on it, fitted with a 6-inch diameter snap-in and out duct, out to the aircraft for the 8:00 P.M. flight and the 5:00 A.M. flight. I would back the truck under the nose of the DC2s and DC3s, open a metal flap, and insert the duct head into the hole. Then I turned up the engine, running the compressor to cool down the planes while they were being serviced and the passengers deplaned or were boarding. It was an easy job. I also checked the oil in the engines. After the refueling and boarding was completed, I would disconnect the duct and drive the truck back to its parking slot. I worked thirty minutes in the morning and thirty minutes in the evening and was paid $50 a month! As a soldier, I was receiving only $21 a month and $13 ration money.

Gradually, I earned my Private First Class stripe, then a 6th Class Technician. A year and a half later, I became a Radioman Second Class Specialist, which paid $72 a month. A Signal Corps officer who flew out from Fort Sam Houston gave me that test. I also had to secure an Arizona driver's license and a Vehicle Driver's License, which was administered by Major Pursley on one of his visits out to check on us.

In what little spare time I had, I worked with a Senior Operator, Harold Harbin, building a civilian amateur transmitter. Harold had a home-built radio receiver with plug in coils for the 10-, 20-, and 40-meter bands. He also had his ham radio operator's license. I passed the exam for that which was given by a radioman from the Federal Communications Commission. My call sign was W6PUH.

Harold and his wife, Jan, lived in a neat adobe house across from the airline terminal. With a Corporal's pay, rations, and quarters, they were able to live above the level of most people in the area.

We needed an outside antenna to transmit long distances on the continuous wave 40-meter band but our DOC (FAA) friends told us we couldn't erect a tower that close to the airport.

Harold and I used a formula from an old radio guidebook on transmitters to get the length of the antenna wire we would need using a matched stubbed lead in to his home near the government fence. There were a lot of salt cedar and scrub bushes along the outside of the fence. The fence itself was composed of five strands of highly galvanized barbed wire strung on dry-treated fence posts and was about five feet tall.

We got all our insulators, wire cutters, and a 6-foot long piece of barbed wire from a scrap heap in the desert. After midnight, on a bright moonlit night, we could see to work without extra lights and we set to work. We cut the top strand of barbed wire, put in an insulator, and spliced in another piece of wire to the insulator fit properly. It was a rough job keeping the tension on the wire, which sounded like a violin string when it was plucked. We wrapped a long rope around the nearest post and I pulled on it to keep the tension.

While we had the wire that was to be our antenna loose, we put in another insulator for the so-called matched stub lead-in. Approximately eighty feet down the fence we put in another strain insulator and drew the barbed wire as taut as we could.

We ran a twisted pair of wire leads in to the transmitter, running along Harold's side of the fence, into the spare bedroom he called his radio room. It really did work when we sent out our CQ's (code for "I'll talk to anyone").

We spent the night Morse Coding radio hams all over the country! They couldn't believe we were using a portion of a barbed wire strand on a fence for an antenna with a forty-watt transmitter we built ourselves! Our receiver was a small Collins Receiver with plug in coils for different bands. It was perfect for what we needed in the desert.

Someone told our Non-Com Officer and the Tucson Airport manager about our antenna. We thought we were in for a lot of trouble for sure! Sergeant Simpson came to Harold's house to look it over. When he saw how neatly it was done and when Harold showed him all the QSL cards (verification from other hams in the USA), he was so surprised and he had a good laugh.

"I don't know if this is misuse of government property," He said, "or defacing the wire fence, but if anyone complains, I'll say it's an experiment for a higher powered back-up system, in case power goes out at the radio shack." He shook our hands and left, still seeming to be puzzled that it actually worked!

Since only three or four of us lived in our barracks, transients could use the extra bunks for fifty cents a night. I was given the job of replacing sheets and laundering the sheets, towels and pillowslips and the money for the rentals was mine to keep. I made a little extra spending money that way because I got a good deal with a laundry service for all of us. We called the place "Hotel-de-Gink", as all were called in the 8th Corps area Airway Detachment.

We had a problem with ants getting into the house so I caught a horned toad about six inches long. He looked like a prehistoric monster but was harmless. He kept the place free of ants and crumbs. His den was a box under the electric range in the kitchen.

One day, a fighter pilot, a Major, stopped over on a flight from Wright Field, Ohio. He was sitting on the bunk with his bare feet on the maroon concrete floor when the horned toad spotted a candle fly near the Major's foot. In hot pursuit, the toad ran across the Major's foot. The Major leaped onto his bunk, screaming, "Kill the damned thing!" He was shaking all over.

I went and picked up the toad and put him back in his box in the kitchen. I had a hard time convincing the major that it was harmless. After that, when we had overnight guests, we made sure to introduce "Little Pancho" to them and to keep him in his box with food. He loved raw hamburger and steak!

It was very sweet and so nice to become a "family" with the American Airline and the Department of Commerce men and their wives. Every month, we military men got a great discount on several cases of beer. We were always known for our ability to "scrounge up" whatever we needed, up to and including "midnight requisitions".

We secured a large number of T-bone (Porterhouse) steaks the same way and other items needed for a party. A few hours before the party, we would put the beer in a washtub packed with ice to cool. The women made coleslaw and potato salad and cooked pinto beans.

After the last American Airline flight left at eight o'clock on a Saturday night, it was party time! A couple of the men had already grilled the steaks and made garlic toast on a grill behind one of the airmen's houses. The couple that ran the restaurant closed up and a few of us would push the nickelodeon out on the covered part of the tile floored patio, which extended beyond the arches. We also moved the restaurant tables and chairs out there, along with the plates and utensils we needed. It was a fantastic party! Some Sunday mornings, we got into bed just before dawn. There were no neighbors to worry about and always great dancing music and good food.

The only thing lacking were girls. But that was solved one day when three student nurses in training at St. Mary's Hospital came out to see the planes coming and going at the airport. They lived in a dorm above their classrooms and one Sunday, a Sister drove them out to the airport. I was coming out of the restaurant, on my way to the barracks, when the Sister stopped me to ask a few questions and we got to talking. She asked if it was lonely living way out here in the desert. I told them how nice it was and about our monthly Saturday night parties.

I noticed the girls looking at each other and smiling. Before they left, they introduced themselves and asked my name.

A couple of weeks later, it was party time again. I told the wives and the other guys about the three nurses who could come to the party. Everyone agreed that I should call them. I did and went to pick them up at 7:00 in the lobby of the hospital and brought them back to the patio. Stan and Fran Roberts from the restaurant treated them all nicely.

Party time was another huge success and I got the nurses in my Chevy Roadster with a rumble seat and top down and back to their dorm by the 11:00 curfew. No one got out of line and no one caused any problems. The nurses seemed to enjoy the parties and a few times, it was past the 11:00 curfew when I got them back to their quarters.

I would park on a side street and they would climb the fire escape, after I jumped up and grabbed the bottom step and pulled it down for them, holding it as one by one they climbed up to the hall window landing. I could never figure out why, when you were supposed to be quiet, that women, after a couple of beers, found everything so funny and giggled so much!

After a year and a half at Tucson, I was going to be transferred to El Paso's Biggs Field. I went by St. Mary's Hospital to say goodbye to the student nurses and to thank the Mother Superior for allowing them to attend our parties.

Mother Superior was very nice and thanked me for my part in getting them there and home safely. After a brief goodbye to the tearful girls, I shook hands with Mother Superior. As I turned to leave, she said, "Joseph, I am so glad you were able to jump high enough to reach the fire escape steps and help the girls up." She had a twinkle in her eye as she continued; "I always left that window unlocked for them so they could get in!" I was stunned, to say the least!

During my first six months at Tucson Airdrome, I was able to help an electrical engineer, O.D. Perkins, who had been hired by the Signal Corps to install four large antenna systems about two miles away from the barracks. The Department of Commerce had five huge towers there for all the aircraft on the beam flying during inclement weather. We had a transmitter building there that resembled a small cottage on a lake, not on the desert.

Mr. Perkins told me he had looked at my 201 file and needed someone who had experience with HAM radio and antennas and who was big and strong because a lot of hard digging and lifting was required, especially the new radio 10-channel, 1000-watt transmitter, etc.

Mr. Perkins had hired two laborers to dig the well-like twelve-foot hole for the 90-foot, eighteen-inch diameter pole. Within two days, the hole was dug, along with a sloping trench leading from the hole for about thirty feet.

The electric utility company rolled in the second day with trucks and cables to set the pole for us. Eight men rolled the pole butt first into the hole. Cables were attached to the pole and two large trucks pulled the pole up while a third truck kept the pole from toppling into the radio transmitter house.

It worked. The pole climbers drove large spikes into the pole, climbed up and attached a cross beam at the top and another about seven feet from the ground to hold the two vertical heavy wire antennas for the 10-channel transmitter.

It covered all aircraft frequencies in the military aircraft with plenty of power to reach them anyplace in the Southwest, either by voice or Morse code. There were two red obstruction lights on the pole, one at the top and one at the middle, both hooked to timed switches, turning them on at dark and off in the morning.

After a lot of cleaning of the three seven-foot tall transmitters with hundreds of relays, switches, etcetera, we were on the air!

After all that was done, we had the utility people from the city bring out four 30-foot poles for our long range War Department Network, called a "rhombic antenna. We dug six-foot holes in the clay under the sand in a diamond-shaped pattern. One pole was near the transmitter building; the opposite pole was 300 feet away, and the other two were set on the other points of the diamond shape. Just like the 90-foot pole, we dug angled trenches, and then used an Army truck and cable to pull the poles up. All went well until, when the pole was almost upright, the steel cable snapped.

The man holding one long rope to steady the pole let the rope go accidentally and the pole swung toward me. I turned the rope I was holding loose and ran through four or five foot tall cactus, called "jumping cactus". This particular cactus had a tendency to sink barbed spines into anything or anyone that brushed against it, break off, and sink even more barbs in because of the weight of the piece.

I had more than thirty cactus needles in my right leg and thigh. I had to remove my coveralls and lay on a cot in the transmitter shack while one of my fellow airmen pulled the spines out with a pair of long nose pliers.

I had a bottle of Tequila and took a couple of long swigs to fortify myself before he started pulling those quills out of me. A large piece of cloth was soaked with Tequila and used to clean the wounds. After that, I got dressed and we got the pole set and began digging twelve inch deep trenches in the shape of a giant wagon wheel radiating out from the transmitter building for the heavy copper ribbon used to provide a good electrical ground for all the transmitters. Because the desert is so dry, we needed these to get out long distance transmissions. It worked.

I took a lot of kidding about running through the cactus. Once, I found a nicely wrapped package on my bunk. I opened it and found a pair of chaps like the cowboys wear in places like that. We also had to carefully check our clothes for scorpions every morning and shake our shoes and socks before putting them on. There were all kinds of poisonous critters out there and you really had to look before you stepped and you never sat on the ground! Rattlesnakes and Gila monsters were also a worry but I never saw many near the buildings.

One fellow told us a story about cowboy who was herding cattle near the Rio Grande in Pecos, Texas. He put on his boots and was stung by a very poisonous scorpion. The foreman sent one of the other men for a doctor, some five or six hours away. Meanwhile, the foreman told the cowboy he'd better start praying! The cowboy had never been inside a church in his life, but said he'd give it a try.

"Oh, God," he prayed, "Please come down from Heaven and help me. Don't send your son, Jesus, because this ain't no boy's job. I've been stung by a scorpion that was made for a purpose but I wish he had not struck me. I did him no harm. If You help me this time, I will never call on You again, 'cause I know you have a lot to do without being bothered by a poor cowboy. Amen."

The desert was very hot during the day—over a hundred degrees in the shade—but there wasn't any shade in the summer, except for a few sycamore trees growing near the dried up river beds, which were a long way from us. While we were doing all the digging, setting the poles and stringing the wires on the poles, I wore a large "coolie" hat with a sliding catch on the cord that went under the chin to hold it on your head. I also wore soft cotton coveralls with long sleeves and rough high top leather shoes. I think OD and I lost several pounds of baby fat over the course of those three months. We were in great shape. We consumed large quantities of iced tea in quart jugs every day. None of us suffered from heat stroke.

Occasionally, on a Saturday evening, we would get the gang together, load up our cars with steaks and beer, and drive up Sabino Canyon for a cookout at one of the cabanas near a dammed up pool. We took my battery-operated radio and charcoaled steaks and swam in the pool and later danced. It was a lot of fun.

During F.D. Roosevelt's administration, the Civilian Conservation Corps (CCC Boys) built a lot of roads, covered picnic tables, dams on streams, etc., all over the Southwest and most still stand today. It was a wonderful program and taught many boys a lot of skills that became their lifelong jobs. Sabino Canyon was a work of art.

The city had a lot of nice places to go for an evening of entertainment so it wasn't a dull place to live.

Charles Horton, from American Airlines terminal, his wife, Ann, and I did a lot of sightseeing on Sundays. We had read Harold Bell Wright's books, "The Mine With an Iron Door" and "Oracle". We took a picnic lunch with us and toured the area where a lot of gold was hidden by prospectors and Indians.

It was a difficult place to explore with boulders larger than automobiles all down the canyon but it was fun looking. Harold Bell Wright came from "up North" at age 24 with a terminal case of tuberculosis—he was given less than six months to live.

He found an abandoned adobe hut in the brush covered desert fringe near the Santa Rosa Mountains near Tucson. He stocked it with food and began repairing it. He found an abandoned burro wandering in the brush and took it back to his hut. He built a brush-covered lean-to near his hut and fenced it in for the burro. He was often seen riding the burro wearing nothing but a loincloth and a dilapidated straw hat.

Much to everyone's surprise, a year later he was in great shape and became a writer and kept adding to his adobe hut. He lived to be 78 years old.

Tucson was swamped with people during the winter months; people who had all sorts of ailments and could no longer stand the rigors of the long, cold winters "up North". But, unknowingly, these people brought their problems along with them—pets, wool clothing, flowerpots, and such, and polluted their own homes. Not even the great clean air of the desert could help them.

I had been corresponding with several ladies on a sort of pen pal basis. One was Matilda Morgan in Virginia. We exchanged friendly letters and cards along with pictures of the areas where we lived and pictures of each other. After several months, I received a letter from Matilda saying that due to the loneliness of the small town where she lived, she had begun dating a co-worker at the factory where she worked and that they were planning to be married soon.

I sent her the nicest congratulations card I could find along with a short note saying that I hoped all her troubles would be small ones.

A few days later, I received another letter from her, telling me that the wedding plans were off—there would not be a wedding. I never corresponded with her again. In all the correspondence we exchanged, she never once mentioned that my childhood sweetheart, Ruthie Horton, was her aunt! About fifty years later, after we were married, I asked Ruthie if she knew a Matilda Morgan. She looked surprised and told me that Matilda is her niece! She showed me pictures of the two of them together at family functions.

Later I kidded Ruthie about how she could have been my aunt instead of my wife! She did not find that amusing at all!

General John J. Pershing, who was retired and living in a suite of rooms at the beautiful El Conquistador Hotel in the outskirts of Tucson, was a common sight at the airport. He had a chauffeur-driven limo and he sat in the back, always dressed nicely in "civvies". We airmen always snapped to attention and saluted as he was driven past on his way to view the planes on the ramp. He always smiled and returned the salute.

He became ill one day at the hotel and was not expected to live. Our radio station was swamped with messages of encouragement for him to get well from all over the country. It was quite a time and a lot of work for the General's son who had to handle all the letters, radio messages, and constant reporter's questions.

Our radio War Department circuits stayed busy relaying messages to the hospital where the General had a small staff and to the crew at the hotel. All the preliminary funeral arrangements had been made and it was a very lengthy radiogram, which I will not divulge here.

But the General was a "tough old bird" and recovered, much to our relief and joy. The telephone kept ringing for days with calls from reporters wanting information about the General. We radio operators knew that any information received that was not addresses directly to us could not be divulged to anyone except the person it was addressed to. Any violation could mean, if convicted, a ten thousand dollar fine or ten years in prison or both! We kept our mouths shut about what we copied in Morse code, even on civilian news services, which we did for code speed improvement. After all this was over, it was good to be back in the routine at our detachment.

We were in the flight office getting briefed on new regs concerning the detached service when Lt. Commander Wallace Berry flew in from North Island Naval Base in California. We were all excited about seeing our favorite actor and didn't even know that he was in the Naval Air Reserves! He had to keep up his proficiency by monthly cross-country flight.

When the excitement subsided, a beautiful young lady, a reporter, knocked on the office door and asked to interview Wallace Berry. He was very polite and she sat down across the desk from him and got out her notebook.

Betty, the reporter: "How did you start working in your life as a young man?"

Mr. Berry: "I got a job."

Betty: "Doing what?"

Mr. Berry: "Working as a swamper in a saloon and cleaning bathrooms in a Kansas City whorehouse!"

Betty: "I'll skip that! What else did you do, Sir?"

Mr. Berry: "I learned to lay the piano and got a better job with better pay."

Betty: "How wonderful! Where was that, Sir?"

Mr. Berry: "In the same whorehouse!"

Betty: "I think that will be all. Good-bye and good luck, Sir."

Betty left, got into her car and sped away. Mr. Berry said, "My whole life's history has been made public for years but there's always someone looking for a dark spot in my character!" We all liked him very much and his movies, too.

I learned a lot in the short time—less than two years—that I was in Arizona. Major Pursley flew out on his periodic inspection and checkup of our detachment. He came in late on a Saturday evening, about 6:00 P.M., so he stayed at the "Hotel de Gink" with us. After he had rested a while, I asked him if he had any civvies with him. He said he did and asked why. I told him we had a great place in town, The Desert Shores, and that they had the very best grilled T-bone steaks, good music, and floorshows at reasonable prices. "Let's go as soon as I get changed!" He replied.

After we had changed, we got into my Chevy Roadster, put the top down, and away we went. It was a lovely cool evening. When we arrived at The Desert Shores, a waiter I knew greeted us. He gave us a table out on the patio near the lake's shore but not far from the band. The Major really enjoyed the place. The people at the tables near us were friendly and everyone was having a good time. The food was delicious and plentiful. I made sure that we were given separate checks. Army regulations clearly spelled out that enlisted men and officers were not allowed to fraternize because it could cause familiarity and breed contempt. Major Pursley smiled and nodded his head. "Where did you learn to be so polite and such a gentleman, Diviak?"

"My parents were reared in Europe, "I replied. "And they impressed that upon us at an early age. We also had a Boy's Social Club in high school, where the football coach conducted an off-season weekly one-hour lesson on social behavior, including how to eat at the table!"

"I know a lot of people who should have had that training!" The Major laughed. "I've enjoyed this very much and will have a nice remembrance to tell Mrs. Pursley when I return home."

During dinner, the Major asked about Mr. O.D. Perkins being there and how I liked him. I was completely honest and told the Major that he was the best and that it was a real pleasure working with a man who knew what he was doing. I said how Mr. Perkins was always polite and nice. I said that I had learned a lot from him.

Major Pursley told me that it was great that I felt that way because Mr. Perkins said about the same things about me!

CHAPTER ELEVEN

Change of Station to Biggs Field, El Paso, Texas

Some of the radio operators in our detachment weren't planning to re-enlist but quickly changed their minds due to all the trouble in Europe in 1937 through 1939. We would have a surplus at Tucson and the last man in would be the first transferred out.

I had swapped my Chevy for a Willis sedan in fairly new condition. I was ready when, a few weeks later, I received orders to transfer to Biggs Field in El Paso. Biggs Field was located at the edge of Fort Bliss and was the home of the famous Seventh Calvary Division.

The Indians had slaughtered this Calvary Division of the Little Big Horn fame in Montana, commanded by General Custer. In all his reputed brilliance and built-up press reputation, it was difficult to understand why he left two Gatling guns (forerunners of the machine gun) behind as he marched north to battle with the Indians. Even today, we have Gatling-type guns in some of the B-52 aircraft tail sections. He must have been in a hurry and wanted to travel light. And yet, he had chuck wagons to drag along on the trip. It's a puzzling thing.

I packed up my barracks bag and suitcase and loaded them in my car, along with some gallon jugs of water and some sandwiches and started for El Paso. I stopped in Lordsburg, New Mexico and had lunch with a couple of acquaintances in the detachment there, then continued on to Biggs Field.

All unmarried Air Corps personnel, about a dozen of them, had rooms built along the walls inside a large balloon hanger. This hanger was a remnant of WWI and was used to house a large oblong-shaped captive balloon for the Army units at Fort Bless until 1930.

The huge 90-foot tall sliding doors had been slid open permanently, leaving the 100-foot wide hanger open. Transient planes were parked in there at night and there were a few rooms with bunks for a "Hotel de Gink". There was also a café managed by a civilian couple. Next to that hanger was the Coast Guard hanger and plane used to patrol the boarder.

Our Commander was Major John Kerksey, a man in his late forties. A 0-47 observation plane was parked in the balloon hanger for his use. This plane was well ahead of its time and all metal. I was surprised to see it there. The airmen greeted me warmly because there was a lot of work to be done and not enough people to do it.

Major Kerksey and the NCO in charge and Master Sergeant Hurley D. Jones all made me feel at home. I was assigned to a room with a man named Bob Guice and we became friends. Lewis D. Howell, a classmate from Chanute Field, and an airplane mechanic, Herman Schultz, who had been in boot camp with us were also stationed there.

It had been raining for three or four days and I began to hear whispers, "We can't do it because of the rain." I found out later that they always had an initiation for every new member to the detachment!

I fit in nicely and spent most of my time pre-flighting radios on Air Corps, Navy, and Marine Corps planes that stayed overnight and helping the mechanic with electrical problems. There was very little radio operation there because the Tucson long-range transmitter skipped over us and went directly to Ft. Sam in San Antonio, then to DC.

The first weekend after I arrived, my buddies and I drove to Juarez, Mexico, about six miles from Biggs Field, through El Paso. We went to a nightclub, ate Mexican food, drank Carta Blanka beer, and watched the floorshow, all for $2.00 each! It was real nice, too!

The next day, I received a message from the CO to come to his office. He asked me if I was happy there and I replied that it was wonderful and I enjoyed it very much. Then he handed me a copy of a radiogram from Major Pursley's personnel officer. I had been promoted to Radioman First Class!

I would now receive $82.00 a month, plus $13 for rations, which made me very happy indeed! He told me that any time I had a problem that Sgt. Jones couldn't resolve that I should come in and see him.

The first Sunday morning, I got up a 5:00 A.M. to preflight and check radios with the control tower. I pushed the pursuit planes, one at a time, out of the hanger, then jumped in before they could roll down the ramp. I put on the left foot brake to be sure it was set at a right angle to the hanger, then I started the engine and called the control tower for a radio check. Once the control tower assured me that all was okay, I taxied the plane to the flight line and filled out the page in the "Form One" book as we had been taught in radio school at Chanute.

Just as I finished the third plane, Sgt. Jones came running across the field, waving his arms at me, and almost foaming at the mouth. "You are not on the list to preflight check a plane at Biggs Field or taxi one either!" He yelled. "Why did you do it?"

"No one told me about any list," I replied. "I was trained to do this at Chanute Field!"

"Only airplane mechanics can do that!" Sgt. Jones snapped. "And I don't appreciate you taking over their duties!"

I turned and walked into the café and ordered breakfast. He followed me inside, still grumbling that I was on his "list". I looked him right in the eye and said, "Do people here know how you got that scar on your right cheek, Sergeant Jones? I stopped by the Airdrome in Lordsburg and had a nice lunch with the men there!"

The sergeant turned white as a sheet and started stuttering. "It's a lie!"

"What's a lie, Sergeant?" I asked.

"Skip it!" He snapped. He left me alone.

I was outside the café sweeping out the sand that had blown in during the night when Major Kerksey came by. I saluted him. "Good morning, Sir."

He returned my salute and replied, "Good morning!"

I asked him how I went about getting on the list to do the radio check in a plane's preflight after they had been here overnight. "Easy," He said. "I'll have the clerk do it. I know you're trained to do that. I've read an outline of the courses at Chanute before I sent my men there."

I saluted him again. "Thank you, Sir!" I walked away with a big grin on my face instead of the mean look I'd had before.

When Sgt. Jones went into his office, Major Kerksey told him, "Sergeant, I've been looking over Private First Class Diviak's records and see that he holds a Radio Air Mechanic Radioman's First Class rating and has been trained to preflight aircraft, so I've had Corporal Traier include his name on the list for preflight aircraft!"

Sergeant Jones replied, "Yes, Sir."

Corporal Traier later told me, "I wish you could have seen Jones' face when the Major told him about you!" He said that Jones had later asked him if I'd been in the Major's office. Cpl. Traier told him, "Diviak knows better than to do that! He knows the 600 series of Army regulations and he knows that he should go to you to see the Commander. He also knows that you can't stop him, according to regulations!"

I had Jones worried. After that, he was very cordial and polite to me but kept looking around to see where I was! In those days, the Army had men scattered throughout the ranks who were NOT members of the Army Secret Service but who made reports on anyone who violated regulations, stealing, money lending, etc. The person, selected by one sergeant, was low ranking enlisted man in each organization, was required to make a report each month. The Sergeant gave each man a number and a post office box number in town. Even the Commander didn't know who these men were. I think Sergeant Jones thought I was one of them!

At one time, Sgt. Jones was Non-Com in Charge at the Lordsburg, New Mexico detachment. He was so arrogant and demanding that the men under his command led a dog's life. One day, in his office inside the hanger, he made a mistake and called a Cuban-American sergeant a foul name. The two men began to fight. Sergeant Jones tripped over a chair and fell against an armchair, cutting his cheek. He bled quite a bit but a civilian doctor sewed up the cut.

Sergeant Jones called Sergeant Rodriguez into his office and told him he was going to have him court-martialed for assaulting a superior. Rodriguez just laughed and pulled out a packet of pictures he had taken of Sgt. Jones backing his car up to a fuel pit on the flight line early in the morning. The pictures clearly showed Jones filling up his private car with gas. You could even read the license plate number! That ended the threat. And when I asked about the scar on his face, he knew that I knew about him stealing gasoline earmarked for maintenance while on base. Airplanes at that time used a higher grade of gasoline.

To make Sgt. Jones even more suspicious, I carried a small vest pocket notebook and, from time to time, when he was around, I made notes in five letter groups, like those I copied at the Secret Intercept Radio Station at Dodd Field in San Antonio.

An incident occurred that served to "confirm" Sgt. Jones' suspicion of me. One day, I was working on the flight line fuel pits, which had to be checked to make sure the many-stranded steel and copper wire, about the diameter of a shoestring, had good brass clamps on it so we could ground the aircraft at the pits when we were refueling.

The desert country is hot and dry and has a lot of static electricity. One spark could destroy an aircraft while refueling.

A plane landed and taxied up to my pit. The pilot climbed out and came over to me. "How are you, Diviak?" He said warmly. "You look great, fellow!" I saluted and he returned the salute, then we shook hands.

The pilot was Major Earl G. Twineing from the Army's Inspecting General's office in Washington, DC. I had taken him to the hotels in town a couple of times back in Tucson when he stayed overnight.

Old Sergeant Jones was standing in the shade of the hanger, taking this all in. The Major was on his way to March and Hamilton Fields in California. We stood and talked for a while, then the Major went to the café in the hanger to get some food. When he came back, I had his plane refueled and ready to go.

I saw him several times after that. I really liked the man. My way of selecting the right man for a work buddy by handshake back in the coalmines saved my life several times and it helped now, too. I knew instinctively who to trust and who not to get involved with—people with very negative attitudes, the "bitchers", and potential troublemakers. Old Sergeant Booth, who retired after thirty years of service, gave this advice to us young recruits: "Always keep your nose clean, a smile on your face and, remember, you can be judged by the company you keep!"

CHAPTER TWELVE

The Beginning of a Romance

One Sunday in October, we held an "open house day" for civilians from El Paso to come out and see the planes up close. We had all three Service's planes there and guides to escort the people. Let the taxpayers see what they were paying for!

I was charge of quarters that day and took several groups through. It was quite a success. The last group was only three people, a man named Ed Rogers, who worked for the border patrol, his girlfriend, Ellen Flynn, a schoolteacher, and Ellen's friend, LaRue Cummins, also a schoolteacher. Ellen and LaRue taught at an elementary school in Tornillo, about fifteen miles south of El Paso.

LaRue and I liked each other immediately. As they were leaving, I shook hands with them and was told later I was a lot more cheerful than usual!

Nona La Rue Cummins
Corsicina, Texas October, 1938
Graduated from North Texas University
Denton,Texas
Teacher at Tornillio, Texas,
near El Paso, Texas

Joseph T. Diviak Caretta, West Virginia October, 1937
Graduated from Radio Operators School Chanute Field, Illinois
Piture taken in San Antonio, Texas at Fort Sam Houston.
We met in October, 1939 in El Paso, Texas and were married October 15, 1940.

The following Tuesday was my day off since I had worked on Sunday. In the afternoon, I drove to Tornillo. It was a very small town with a cotton gin, a general merchandise store, and a service station. The schoolhouse where all eight grades were taught was a nice Spanish-style building. There were farms and ranches all around the town.

I pulled up to the school just as LaRue was coming outside for the 2 o'clock recess. I got out and went over to her (I was in civvies). LaRue was shocked to see me—later she admitted that she was secretly hoping I would come to see her. I asked her if she would like to visit with me at the store after school let out at 4 o'clock. She said she would like that.

I left and drove past the ranch where LaRue was staying, then to the town of Fabens, which was larger than Tornillo. I found a café and had some coffee. I also picked up some great roast beef sandwiches and two slices of pecan pie. I sat in my car under a shade tree, watching the nearby Rio Grande. The place was beautiful.

Just before four, I drove back to the general store in Tornillo and went inside. LaRue came in shortly after that, all smiles. I got a couple of Cokes and we drove about three hundred yards to the riverbank and talked and ate our sandwiches and pie. Then we drove to the ranch house where she was living. We sat in lawn chairs in the back yard and talked more and more.

I knew she had a lot of work to do. Not only did she teach third and fourth grades, she also was the band and music teacher at the junior high school.

We made a date for the next Saturday. I would pick her up at 6 o'clock and we would drive back to El Paso and then across the river to Juarez, which seemed like an extension of El Paso.

I was on cloud nine as I drove back to Biggs Field! At last, I had met someone just like I had dreamed of all those years and was having my first date! It was a wonderful feeling knowing that I was finally free to look for real happiness and a future.

My buddies kidded me all the next day, making comments to each other, like "I wonder if Joe is sick? He's so quiet!" I could hardly wait for Saturday to come but being busy helped make the time pass quickly. We overhauled some transmitter equipment in the radio transmitter building, the flight line, and cleaned sand out of the hanger.

On Saturday, I dressed in my best summer clothes. Even though it was October, the temperature was between eighty and eighty-five degrees, but it wasn't too warm for a jacket because there was no humidity. I stopped at a florist and picked up a small but pretty orchid then drove to the ranch, arriving at exactly 6 o'clock.

The maid and cook, Rosetta, showed me into the living room and a short time later, LaRue walked in, dressed like a beauty queen! We shook hands and I pinned the orchid on her dress and away we went to Juarez for dinner at the Old Mexico Restaurant.

We each had a Tom Collins and then a wonderful Mexican dinner. A Mexican band played both Mexican and American songs like, "La Galdendrina", "Star Dust", and "Smoke gets in Your Eyes". As we dined, the music picked up and in an hour, the floorshow began, a wonderful mix of hot dances and fast music. After we finished our dinner, we danced for a while. It was out of this world!

People kept looking at us and it bothered me a little. I excused myself to go to the men's room and on the way back, I asked the waiter why people kept looking at us. He laughed and replied, "You are a handsome couple and your date looks like a movie star but no one can recall her name."

That really did it and later, I told LaRue what he had said. She laughed and said, "What a sweet and generous compliment!"

My re-enlistment time was approaching and I had to decide whether I wanted to stay in another three years. It was an easy decision. Major Kerksey swore me in for another enlistment. I received a re-enlistment bonus along with other money due for unused furloughs, etc. I immediately went to the jewelry store in El Paso and bought a single diamond engagement ring, after only knowing LaRue for two months.

We had dinner at a nice restaurant in El Paso and danced and talked and had a great time together. I drove her home and we sat in the car, the moonlight shining so brightly you could read a newspaper. I gave LaRue the ring, slipping it on her finger. She burst into tears of joy, her head on my shoulder. She was really surprised and speechless!

We went into the large adobe and stucco house by the back door that led into a huge kitchen where the ranch hands usually had their meals. They had all gone back to the bunkhouse for the night but Rosetta was there. LaRue showed her the ring and Rosetta hugged us both and cried. Then Mr. And Mrs. Shaffer came in and congratulated us. It was an exciting time.

I finally made it back to my room at 11 o'clock to get a little sleep before pre-flighting a few Navy planes sitting in the hanger overnight.

LaRue was born on a farm near the town of Roan, Texas, which was a very small place. The circuit doctor made a string of house calls from farm to farm, visiting people. He usually had medication with him or, if not, prescribed for people who were ill. He delivered LaRue and when he had completed his circuit three or four days later, he recorded it in Corsicana.

The doctor was known to partake a little of the "spirits" offered to him by farmers along his route. Instead of filling out LaRue's birth certificate with the name "Mary Elizabeth Cummins" as her parents, Nona Halloway Cummins and

Hugh Cummins wanted, he wrote “Nona LaRue Cummins”. They never changed it.

When LaRue was four or five, the family moved to Corsicana and Mr. Cummins went to work at the Southern Ice and Refrigeration Company. They rented a nice large house in a good neighborhood. Five years later, LaRue’s brother, Eugene, was born and Donald five years after that.

LaRue began taking piano lessons when she was eight years old and really loved it. Her mother, Nona, had taught school for a couple of years before she was married. She got her teaching certification by attending a Normal Teacher’s School for a year.

Nona had a lot of drive and pushed LaRue hard every minute of the day. She sent her to the grocery store for sewing thread or needles or something else at least twice a day.

LaRue played the piano at recitals held at the church and at services in Sunday school. People loved to hear her play; she seemed to have “the right touch” on the piano.

She didn’t date much and graduated from high school. She had worked as a clerk in a store in town and taught piano.

Mr. Cummins left the Ice Company and, with a partner, began buying and selling cattle, hogs, and horses. They did very well because he could guess the weight of the cattle and hogs and knew what they would bring on the market. He had a very good reputation in all his dealings and his handshake on a deal was as good as a written contract.

In 1936, LaRue enrolled in North Texas State College (now North Texas University) and received her Bachelor of Arts in three and a half years. She was certified as a music teacher and received her State Teaching Certificate for all grades from first through twelfth. Her first assignment came after Christmas (1939) in Sun Ray, up in the northern panhandle of Texas, which was all flat country, very hot in the summer and very cold in the winter. The area was plagued by a lot of dust storms, too.

LaRue enjoyed teaching several elementary classes but the frequent letters from her concerned mother to Mr. Robinson, the school superintendent angered LaRue. Mr. Robinson was very nice and told her not to worry, that he would take care of it. He wrote a letter to her parents, which stopped the letters to him but not to LaRue. She survived, however, and applied for and got the teaching job in Tornillo. She had wanted to get as far from Corsicana, which was about sixty miles south of Dallas, as possible. El Paso was about 700 miles west of Corsicana.

El Paso was a great place to live. They not only had movie theaters, they had Legion Hall, where a lot of the top theatrical companies, orchestras, and famous operas were performed. There were a lot of good restaurants and dinner theaters and, of course, a lot more entertainment across the river in Mexico.

The "CCC Boys" had built roads, parks, cabanas, barbeque pits, and pools in McCallegon Canyon just outside of El Paso. There was also a football team back then, called "Texas School of Mines" that played in the Sun Bowl on New Year's Day.

We took trips all around there and into a small town near Juarez. The large cathedral there was a beautiful work of art. I marveled at how it was possible, in the 18th century, for the Mexicans and Indians to haul the 16-inch square, 60-foot long mahogany beams that supported the roof and ceiling of the church. There were twenty of these beams. It was at least 650 miles to the Pacific Ocean and the same distance to the Gulf of Mexico, and the Rio Grande was not navigable.

One merchant in town told us that slaves or peons and Indians carried the beams overland. They used cross poles lashed to the timbers with several men under each pole on each side of the beam. Wagons could not be used to transport the beams because there were no roads, only trails.

The best view of El Paso was from the Scenic Drive around the mountain. The road had guardrails along the outer edge and there were a number of nice parking areas with views of the city. It was especially beautiful at night and very cool there, even in summer, since it was 3,900 feet above sea level.

At that time, the crime level was low. When you drove into Mexico, you gave a kid a quarter to watch your car, which was a wise thing to do! I never had a problem. The police were also nice, on both sides of the border. As my old Sarge said, "Keep your nose clean and mind your own business and you'll be fine!"

We could bring certain things across the border in small quantities duty-free, even a quart of Tequila. Leading brands of perfume were half the cost as in the USA and one bottle was duty-free. Our Christmas and birthday shopping was planned months. We carried a shopping list even if we were only going there for dinner.

Some stupid people tried to hide things, like perfume, in their spare tire. The tire would hold a lot of bottles and looked perfectly normal when air was put back in it. What they didn't know was that the merchant would call the inspection station on the US side of the border and give a full description of the person and their vehicle!

I was told there was a payoff for this information, which is probably true. Why else would the police let car after car go by after only a brief inspection, then pull one car out of line, direct it to a parking area, then almost dismantle it until they found the contraband? The inspectors knew every trick in the book!

White Sands, New Mexico, Heuco Tanks, a beautiful rock formation and caves and lots of churches were nice places to visit and enjoy. But there weren't many places to have a picnic except in the canyon and Hueco Tanks, because there were no trees out in the desert or along the highways.

The 21st Observation Squadron from Ft. Riley, Kansas flew down to stay with us. They used their 0-47 planes to photograph and map the border and both sides of the Rio Grande. They stayed with us a couple of months and right off, two of their airmen landed in a Juarez jail! It took a week of negotiations between consuls and other officials but they were finally sent back. All they had on was a pair of slacks. They were changed men after that and we had no more problems.

LaRue and I spent a lot of time together in the evenings and on my days off when there was no school. Sometimes some of her teacher friends would come with us on a Saturday night. One day, I arrived at her school a bit early. LaRue was working with a student, Wendell Harris, who was having difficulty learning to find the lowest common denominator in fractions. LaRue had been taught one way to do it and I had been taught another. When LaRue was called to the office, I worked with Wendell, showing him the way I did it and he learned quickly. LaRue was quite surprised!

She asked my why I looked so strangely, as though I were in a daze. I told her about a dream I had a couple of years earlier, while at Fort Sam Houston. In the dream I met a beautiful lady teacher and she was having a problem teaching a male student and that I helped. The boy in the dream became an "A" student because of my friendship.

"It is a nice dream!" LaRue said. After several months, Wendell did change and she was so proud of him and so were his parents. Yes, he became an "A" student!

One night, after LaRue and I had dinner, I took her home early because she had a heavy schedule of report cards to turn in. I headed back to Biggs Field. On a curve on a back road, suddenly my car's front wheels turned outward at different angles. The tie rod had snapped as I went around a flat curve! I hit a concrete wall at 45 miles an hour!

The engine was pushed back against the seat to my right and I had twisted the steering wheel off, but I only had a few bruises because I was wearing a heavy leather flight jacket. I managed to get out of the car and limped a few hundred yards to a service station where I called my friend Lewis Howell. Lewis came to get me and called a tow truck.

When we got back to the barracks, I showered and went to bed. In the morning, a flight surgeon from Fort Riley Squadron checked me over and sent me to William Beaumont Hospital for a complete checkup. I had a badly bruised right knee. I was kept immobile to prevent "water on the knee".

I sent a telegram to LaRue explaining that I'd had a slight accident and wouldn't see her for a few days. I assured her I was fine, that there was nothing seriously wrong with me.

That evening, I heard a commotion and in came LaRue looking very worried and a bit angry. The lady nurse tried to keep her out, just to show her who was in charge, but LaRue would have none of that! Most of the men in the ward were "CCC Boys" who had broken bones from cutting down trees and hauling rocks in the Grand Canyon area.

LaRue was quite relieved but still a little angry about the telegram saying "a slight accident"! She was beautiful even with fire in her eyes! I knew then that I had chosen the right lady for a sweetheart. She brought me a book and a nice "get well" card.

LaRue checked on my car and I gave her a check to pay the mechanic. She drove the car up to see me and said that the mechanic had said that the tie rod (not used these days!) was rusty at the break, which meant that the car had been driven over a high object. I never loaned my car to my fellow soldiers after that.

I was released from the hospital after a couple of weeks wearing a half cast from my ankle to my rump. It looked like a long trough and my leg laid in it and was secured by elastic tape strips. I had to use crutches and was marked for light duty.

Sgt. Jones had me working the flight line within a few days, using only one crutch. I was refueling planes and repairing cables at the pits. In three weeks, I was to go back to William Beaumont Hospital to see Major Scott for a checkup.

Each night, I had a new blister under the top and bottom of the cast. I would cut off an inch or so at both ends of the cast with my pocketknife so I could work the next day. I even found a cane in the supply room that I could use instead of the crutches.

When I went back for my checkup, Major Scott was furious. There was only about a foot of cast left out of the 30 or so inches of the original! He removed the cast and told me to walk down the hall. I did. Then he told me to run toward him. I did. He was still annoyed and I asked him why.

"According to the staff here, you would have a slight limp for life!" He said. Then he asked if it had been painful. I told him only for about twenty or thirty minutes in the mornings and evenings.

Major Scott released me to duty and said, "It may have been your continued use of the knee working on the flight line that did it!" He called Major Kersey and told him about it, then asked who had put me on the flight line. "Master Sergeant Jones!" Major Kersey said. They talked about it a while. Then Major Kersey called Sgt. Jones into his office. The Sergeant was asked to give his interpretation of what "light duty" meant. He was surprised to see the word "light"!

The Major instructed him to read those comments on the reports more carefully and dismissed him with a smile. Sgt.

Jones came to me and apologized. I told him to forget it — no harm done!

A few days later, I received orders to transfer to The Reserved Flight Squadron at Brooks Field in San Antonio, where, much to my surprise, I would stay before being sent to Instrument Flying Instructors School at Chanute Field, Illinois. What a wonderful surprise! I'd always had a feeling that someday Major Pursley was going to give me a big break and this was it!

I had read and studied a lot about instrument flying since I had helped Mr. Pratt with the radio beacon tower transmitter equipment in Tucson. I learned a lot about radio navigation and direction finding equipment. "Someone" was really looking out for my welfare!

One evening, I helped a fellow radio operator take late messages on the phone. When we were through, we were leaving and Major Kerksey asked me to step into his office. We sat down for a good old fashioned "visit". "Diviak," Major Kerksey said, "I was going over your 201 file, updating it, and I noticed you are from West Virginia. Have you ever been in Boone County?"

"Yes, sir," I replied. "We lived there in 1921 but left after the strike."

"Where did you live in reference to the school building?" He asked.

"Diagonally across the street, on the corner of the road that runs along Little Coal River!" I said.

Major Kerksey became a little nervous and he stared at me intently. "Do you remember the Federal Militia being there?"

"Yes, sir," I said. "I was about six years old. I would stand on a railing on a white wire fence and watched as the Army men stacked their rifles in a teepee formation and set up their tents." I smiled. "One young man with shiny boots came over and asked if we had a well. I opened the gate and took him around back to our well. My mother came out, said hello to him, and told him they could have all the water they needed!"

There was a long silence and finally Major Kerksey said, "Joe, I was that Officer in charge of the Militia sent there to keep order. It's unbelievable and fantastic that we should meet again after eighteen years!

"I went through pilot training and got my wings. I've been all over the country since then and here we are, together! Unbelievable!"

I told the Major about the strike and about the two gunmen the Coal Company had hired and how they disappeared after they were tried in our little schoolhouse across the street from our house.

We talked about how I had later worked in the coalmines and became the recording secretary for our Local of the United Mine Workers Union. I told him how I'd had the pleasure of meeting John L. Lewis, our Union leader.

Major Kerksey sat there in amazement. He said he wished he had known who I was earlier—we could have enjoyed cookouts at his home in El Paso and that he would love to hear more about the coalmines!

"Joe, you have sure earned your spurs! I knew you wouldn't be here long because Major Pursley, our Commander at Headquarters, has a very high regard for you and your welfare. If you ever need anything—a recommendation or a friend—be sure to call me!"

I thanked him and left.

CHAPTER THIRTEEN

Transfer to Brooks Field, Texas

I packed my military necessities in one bag and civilian clothes in another, which would go into the trunk of my car. LaRue would have transportation back to Corsicana via San Antonio where I would meet her. Her school would be in session through June 1st and I had to leave the first of April.

All the arrangements had been made. Captain Bailey flew in on April 1st in a new BC-1 (basic trainer) and landed at Biggs Field to pick me up. We had lunch in the café in the hanger, then I said my goodbyes to everyone and thanked Major Kerksey and Sergeant Jones and we left.

We took off in the basic trainer, which had a cruising speed of 160 miles an hour. It was 500 miles to Brooks Field in San Antonio—a direct flight. Our hanger was Number Eight and next to it was an old air traffic control tower from when it had been used as a training base. That was home for six of us.

We could see for miles around. We had roll-up shades to keep the sun out. It was cool and nice up there.

I met the rest of the gang of about fifteen men, and then later met the fifteen or sixteen pilots who were in the Reserves, getting their training. What a beautiful setup! We "messed" (ate) with the 22nd Observation Squadron–the Squadron who made the movie "Wings".

It was nice to go to the mess hall, grab a tray and utensils and have pancakes, sausages, potatoes, eggs, or whatever was on the fare for that morning, and toast and coffee. Lunch and supper were the same way.

I had twelve planes in the hanger to check and was given an assistant, Bob Bailes from Monroe, Louisiana, to train. We all got along just fine. There were no problems.

I kept writing to LaRue and she to me. The first of June, she drove from El Paso to San Antonio to see me. I got her a room at a tourist court near Brooks Field for the night. I had a three-day pass so that I could drive back to Corsicana with her and bring my own car back and meet her parents.

We talked late into the night, then I returned to the base. I checked out early the next morning, packed LaRue up, and we headed for Corsicana, some five or six hours away. There were no interstates back then!

When we arrived in Corsicana, we went to her house—a large nice home on a good street in a nice neighborhood. I met her father, then her two brothers, and finally, her mother. I knew I was in trouble the minute I shook her mother's hand and looked into her eyes. It was a short three-day visit, but during that time, I met a lot of LaRue's aunts and uncles and cousins.

LaRue was a bit upset at her mother's attitude toward me, which stemmed from the fact that I was a foreigner, a Catholic, and a soldier–three strikes and you're out!

I worked hard at Brooks and drove the five or six hours to see and spend time with LaRue. LaRue, her father, her brother Don, and I would go fishing on the pier at the city lake and sometimes in the numerous ponds on nearby farms. We caught fish and had a great time, but in the background I could hear LaRue's mother, Nona's, voice—she didn't like me but LaRue's father did.

One day, I asked LaRue what her mother's plans for her were before I came into her life. She told me that her mother always wanted her to be a schoolteacher and teach and live in Corsicana. Then each summer they would travel together to places she had always wanted to see.

"How far can two people travel on a teacher's wages of $80 a month?" I asked LaRue.

She said that she had asked her mother that and her mother had said, "We can manage!"

Captain Henry M. Bailey, our Squadron Commander, called me in his office and told me that in three weeks, I would go to Chanute Field for three months to attend the first class for Instrument Flying Instructor's School and that I would be given Corporal stripes to go with my Radioman's First Class rating in order to keep me out of regular chores while I was attending school.

I had enough sense to know that my entire Army career hinged on passing that course at Chanute Field. I wrote to LaRue, asking her to come to San Antonio to pick up my car, a 1937 Ford with a "V-8" engine—so that she could use it while I was gone.

She and her cousin, Muriel Goodwin, came a week later and stayed at a motel near Brooks Field. LaRue took my car when they left.

I took the train out of San Antonio, bound for Rantoul, Illinois. I was able to see LaRue for a few minutes when the train stopped at Palestine, near Corsicana. It was a good boost to my morale! I arrived in Rantoul the next day and was met by a military vehicle that took me to the barracks that had been added since I had been there in 1936.

It was a hard, full schedule and very strictly regulated for those three months. There were ten trainer simulators in a hanger, which was air conditioned and dust-free. We all had to learn to fly the training simulators. Even though they never left the ground (or the floor!), the flights were very realistic. It didn't take me long to learn to fly by instruments, even with the rough air bumps, etc., that they simulated. Then we learned to fly the beam of the radio directly from one field to another. I learned radio navigation and how to teach others to do it. We also had meteorology classes given by Dr. James Taylor from MIT and classes on how instruments were manufactured, so that we could not only instruct cadets and student officers to fly instruments, but to repair and maintain the simulators with all its radio, vacuum, and mechanical equipment.

Before we graduated, we were taken on a flight in a BC-1 trainer. We sat in the rear cockpit that was covered by a hood, flown away from the base, and told: "Get us home!"

In just a few minutes, I had the beam and code letters and that was all I needed. As the whine of the radio beam in my earphones increased, I knew I was headed in toward the beam towers near the airfield. I was flying at 2,000 feet with my airspeed at 140 mph, then I started to drop down at 300-feet per minute. I picked up the outer marker after I had passed through the "cone of silence". I was 500 feet over the fence. I came in at the end of the runway where I should be for practicing and the pilot took over and landed the plane.

I was the happiest man in the world right then! I could see Staff Sergeant stripes on my arms and see myself married to LaRue. My dream was finally coming true!

After graduation and picture taking, four of us left Chanute bound for Texas in a large Chevy sedan. We drove all that afternoon and all night to get to Dallas, changing drivers every two hours.

LaRue was attending summer classes at a college in Dallas, getting a few credits toward her Masters in Music degree. I called her when we got to Dallas and she picked my up at a service station and we drove to Corsicana. LaRue dropped me at a service station near her house. Her mother was very upset and it was best that I didn't go home with her–it would only add fuel to the fire.

In about twenty minutes, LaRue came speeding back to the service station, brakes squealing as she came to a stop. I could see she must have had a rough time with her mother. I didn't say a word, just climbed quietly into the passenger seat, and we took off like a rocket, headed for San Antonio.

From time to time, I looked into the back seat and saw suitcases on the floor, a floor lamp, a waffle iron, and an electric iron back there. I sat quietly.

After about an hour, LaRue began to tell me what had happened. Her mother told her that she could go, but if she did, she was not welcome back home again! Finally, she said, her mother told her, "If you will stay, I'll give you a church wedding." But LaRue told her mother, "No. That won't happen because you'll find excuses to prevent it!"

As LaRue was getting into the car, her mother screamed, "I know about those foreigners! In a few months, you'll be back with a bunch of kids!"

I laughed when she told me that and said, "I think your mother has overestimated me in that field, but after we're married, we shall see!" We were both laughing when we stopped for lunch in Waco. We went on to New Braunfels, which is about thirty miles north of San Antonio.

We went to city hall.

CHAPTER FOURTEEN
Getting Married

At city hall, we were told that I needed a Wasserman Test before they would issue us a marriage license. We drove around the block to a doctor's office—an ex-Army medical officer–and I explained what I needed.

He took a tube of my blood and told me it would take a day to get the results back. I explained that we wanted to get married that day and that I'd had a blood test just three months ago, just before I left for Chanute Field.

"Oh!" The doctor said. "Are you in the service?"

"Yes, sir." I said. "On my second hitch, stationed at Brooks Field."

"Great!" The doctor replied. "I'll give you a certificate saying you're okay. Just leave me your phone number before you leave so that I can reach you if there is any problem."

I wrote down my number, he gave me the certificate and wished us the best. We went back to city hall for the license, then to a minister at a parsonage near the church. We were married there, with the preacher's wife as a witness. At last, we were together!

We continued on to San Antonio and apartment hunting. We drove across the city, out a few blocks on Roosevelt Avenue and there was a sign—an upstairs furnished apartment for rent, with an outside staircase leading up to it.

We stopped and met the lady who owned the house, who we liked immediately. We told her we were newly weds and needed an apartment. She took us upstairs and showed us the apartment, which consisted of three rooms: a kitchen/dining room and a living room and bedroom. It was well furnished, with dishes and all. I explained that I had pay coming from the Army Air Corps as soon as I checked in. She said that was no problem and that we should pay on the first of the month.

We moved in and went shopping for groceries. We were given instant credit at the Red and White Grocery Store. When we returned to the apartment, there was a small pink and white three-layer cake with a candle in it sitting on the dining room table. It was from our new landlady!

It was a wonderful evening and we couldn't thank the landlady enough.

CHAPTER FIFTEEN

New Assignment—Kelly Air Force Base, San Antonio

The next morning, I dressed in my uniform and checked in at my Squadron Orderly Room. Captain Bailey came out while I was there and congratulated me on my marriage. Then he invited me into his office, saying he wanted to talk to me.

"The plans to establish a Link Trainer Instrument Flying School here have been scrubbed. It will be opened in Hanger Eight at Kelly Air Force Base. You'll be transferred there and the rest of us will come there for our training update on Saturdays. Cadets will fly during the week." It was a pleasant shock and sounded great to me!

When I was transferred to Kelly, we had only one trainer but more were expected. There would be twenty-four in all and Hanger Eight, which was in the second row of hangers off the flight line, would be air conditioned and made dust-free.

Master Sergeant Llewelyn had been trained at the factory in Binghamton, New York and was in charge. I was the only other instructor. Sgt. Llewelyn was a 20-year man and had been a coalminer as a young man in Colorado and we hit it off nicely.

LaRue and I moved to a duplex near the main gate at Kelly Field. It was nice and large. I received my Staff Sergeant stripes in a short time and we began expanding the Department of Instrument Flying School.

We had more instructors and twenty-four trainers. Later, six more were added. I was made Senior Instructor and supervised the other instructors, checking daily to make certain we were following the correct teaching methods. I did a lot of teaching, too, to keep in practice.

Our Squadron was the 12th Air Base Squadron, Major Kinsinger was our Commanding Officer, and Mack Williamson was our First Sergeant. Sgt. Williamson was a nice man who spoke in a soft voice, but with plenty of authority.

While at Brooks Field, LaRue and I had joined a Southern Baptist Church, but in order for me to become a full member, I had to be baptized into the fellowship, which was one of the tenets of the Southern Baptist Convention.

For the first time in my life, I was really nervous, but LaRue said that she would be baptized at the same time. She had been baptized years before, at the age of eight, but she didn't remember much about it.

One Sunday morning, before services, LaRue and I along with two other couples, were baptized in front of the entire congregation. The baptismal was behind the pulpit and choir loft and all you could see was the pastor and prospect from the shoulders up. It was a nice ceremony.

I had been christened by anointing in the Catholic Church when I was a child, now I had been baptized by water as an adult. For years afterward, whenever I was asked if I was a Christian, my reply was, "I think so. I've been dry cleaned and wet washed, so I must be!" Some people didn't think that was funny, but my pastors over the years sure did!

The Church held socials for the young couples and there were Church dinners. LaRue played the piano very often at church services. There were between five and six hundred members in that church and all were nice people, a loving and caring fellowship!

Even after we moved closer to Kelly Field, we continued to attend the Riverside Baptist Church and shop at the friendly Red and White Grocery Store.

Helen was born at Sam Houston Hospital a year after we were married. We had a family at last! Helen was a good, happy baby.

One day, in the Squadron Orderly Room, I was talking to First Sergeant Williamson and he asked if I was going to re-enlist, explaining that a new Army regulation stated that an enlisted man below the rank of Staff Sergeant who married without permission would not be allowed to re-enlist.

Between the two of us, we figured that I had better get written permission from our Commander and get the letter dated prior to the date we were married. I wasn't alone—Sgt. Williamson and others in the Squadron were in the same boat.

Permission was granted and we went to the County Courthouse and got a marriage certificate. We asked the assistant pastor at the church to marry us. He wasn't sure it was legal to do that, but the Provost Sergeant said that it was. Afterward, I submitted the paperwork to my Commander and was allowed to re-enlist!

We had 350 cadets plus a few student officers from other branches of the Army going through advanced training at Kelly Field every three months. We flew cadets for only a half an hour each time because of the crowded schedule, but before they left, they had eighteen hours of flying in the simulators and had passed the test flying in real airborne planes.

Within a year, I was a Technical Sergeant. Another year later, I was a Master Sergeant. Then I took the exam and physical for Warrant Officer in Communications to be the C.O. of an Instrument Flying School.

Letter of commendation from Major Bailey, Major Pursley, and Major Kersey helped a lot.

I had finally made it!

CHAPTER SIXTEEN

Pearl Harbor Attacked and War!

When Pearl Harbor was attacked on Sunday, December 7, 1941, I was at home in my quarters in the housing development near the entrance to Kelly Field. I was called to assembly on base to be briefed.

Classes would be doubled and our instructors increased, but we would stay where we were and in between would receive combat training.

My instructors were an assortment of farmers, schoolteachers, lawyers, sales people, radio operators, etc., but all jelled together nicely to make the school very credible, even with an ex-coalminer in charge!

When I was a sergeant and had quarters on base, I had three buddies who loved to fish and hunt. They kept their tackle and guns in the furnace room in our basement. We would get away on Saturdays and go fishing at Medina Lake and river. We had an extra refrigerator that we turned up high and used it as a freezer to store our fish. We would often have a fish fry at my quarters.

During hunting season, there were doves, ducks, and venison, so we had all the meat we needed. We didn't need ration stamps for meat at all.

LaRue would make potato salad and we'd buy a gallon can of corn at the commissary, then the boys would fry the fish or whatever meat we were having and eight men, one lady, and a baby would have a great time together.

Little Helen would sit between the airmen and they would feed her good lean meat or little pieces of boneless fish. Once a month, we were all one large, happy family. The men really spoiled Helen with toys, candy, cookies, and lots of attention.

Sgt. Jackson and Sgt. Hogan were the best fishermen and hunters among us. Many evenings, they would come from their barracks across the parade ground to sit on our front porch and read or write a letter. They loved the quietness and coolness of the huge duplexes.

A lot of interesting things happened at Kelly Field. One day, as Officer of the Day, I drove around the base and saw a soldier digging a ditch for a cable for the perimeter lighting. I remembered someone saying that he was a concert violinist who had studied under Fritz Chrysler and Spaulding—two of the greatest.

I got out of my jeep and talked to him a while. Then I went to the Base Chaplain's office and told him about Ernest Holmes out there digging ditches. "I know that man!" said the Chaplain.

The Chaplain went to the Personnel Officer and had Holmes transferred to the Chaplain's Office. Holmes organized concerts and special music at the chapel with other soldiers who were musicians.

One day, there was a tap on my office door. "Come in," I said. It was Holmes and he asked permission to speak to me. We had a brief chat, and then he asked how I knew about concert violinists. I told him about my roommate in Tucson, Cresep, who practiced his violin two or three hours every evening. Then I said, "Hard labor will ruin your fingers!" I also told him that my wife, LaRue, was a concert pianist. He thanked me and left.

Our Commander, Colonel Pitts (his real name!), came up regularly to fly the simulator. We were both HAM radio operators and we found that he and I had the same problem: because of the war, we needed a copy of our birth certificate to prove that we were American citizens!

I had already written a letter to the Bureau of Census, both in Charleston and Washington, DC, as well as to the county seat where I was born and the Catholic Church in Berwind, West Virginia. I gave him some possible places he could write, which he did. It took another nine months before we finally got a delayed birth certificate. In my case, the Charleston Courthouse records had been lost in a fire, Welch did not record live births, and so on.

It was a puzzle to both of us since we both had security clearances and were in the service, but that was not good enough for the Federal Communications Commission!

I was given a three-month temporary change of duty status to put five supply rooms in order at Kelly Field's Tent City, where there were 2,500 recruits and five mess halls. When the place opened up, there were "ninety-day wonder" officers plus recalls of officer without any supply or mess hall experience. Then there were the political appointees who hardly knew which hand to salute with or to whom! You could spot them a mile away. They didn't have signed receipts for all the clothes, cots, mattresses, blankets, mattress covers, and a lot of other things that had been given out to soldiers without signatures. What a mess!

I had a couple of buddies assigned to me who were supply sergeants. We tore old blankets and mattress covers in half and did a lot of "scrounging" night and day. There was approximately $50,000.00 worth of property misplaced and not signed for. It couldn't have been taken off the base because all vehicles were searched. But it was missing.

We had the same thing in the Orderly rooms—morning reports, personnel files, etc., were all in a terrible mess. Mr. Wood, a fellow Warrant Officer in administration, was given the job of cleaning that mess up. Within three months, we had it all back in shape. Major Wilson in Base Supply signed my release reluctantly but he had no choice, because I had done my job. He kept asking, "How did you do it?"

One day, after inspecting the Command's training facilities, Colonel Hayes, the Inspector General for the 8th Corps Area Training Command, called my commander. He asked him to send Mr. Diviak to the Independence Air Base in Independence, Kansas, on the recommendation of Lt. Co. Bailey, my former boss who was now the personnel officer for the Training Command. They had eighteen Instrument Flying Trainer Simulators on line, all new from the factory, none of which was working!

I was sent there on temporary duty status for three weeks to correct the problem. After an overnight trip on the train, I arrived at Independence on a very cold day, in deep snow.

I checked in with the Base Commander and met the officer in charge of the Trainer department. He was a nice guy but fresh out of the "ninety-day wonder" school! He had absolutely no knowledge of trainers.

I was assigned a room at the Visiting Officer's Quarters and changed into my old enlisted man's coveralls minus stripes, picked up my old troubleshooting volt-ohm-meter and went to work.

I looked over one of the units and discovered that, in spite of the "handle with care" signs on the huge plywood crates, the trainer had been dropped. A capacitor in the radio chassis had broken loose in the power supply. The rest of the units had the same problem, as well as a few broken vacuum tubes.

I made a list of materials and went to the Quarter Master for a voucher, then drove to a radio shop in town and purchased all the parts I needed.

Within two weeks, I had made all the repairs and had the Lieutenant in Charge, with his senior Sergeant following us around, while I screened the crew. I found two former radiomen and showed them what the problem had been. Then we had to test fly all the trainers. The Commander,

Colonel Righter, was very happy! I was treated to a steak dinner and drinks at the Officer's Club the night before I left. He also sent a letter of recommendation to Headquarters at Randolph Field and to my Commander at Kelly Field.

I came back to find my Flying Instrument Department in disarray! The arrogant "Ninety-Day Wonder" who'd been in charge while I was gone had everyone upset. I immediately called in my "Chiefs" (Sergeants) and they told me what had happened.

I went to see my friend Colonel Pitts. He listened patiently while I told him and said, "I'll take care of it."

That officer was gone in a few days and another officer, who had been to Trainer School, was assigned to help me. He quickly learned how to manage training and complete all the paperwork involved.

One of my sergeants was from Coffeyville, Kansas and we were great friends. When he learned that I was being sent to Coffeyville Air Base as C.O. of the Instrument Flying School, he asked if he could go with me. I asked and he was sent there too.

I knew it would be tough because I had a great job and was teaching instrument flying every Tuesday and Thursday evening to CAP pilots and others under the National Youth Program at Kelly Field. The Chief was Mr. Lyndon B. Johnson, the Texas State Administrator.

I was receiving $96.00 a month under as temporary teacher's certificate from the University of Texas. That was going to be eliminated because of the war. I had a "bird's nest on the ground" when I was a Master Sergeant and was receiving flying pay, which was half of my base pay, plus the $96.00 a month for teaching, plus quarters and ration money.

The Quartermaster people hauled all our household goods, including the dishes, to the warehouse where they packed everything very securely, ready to ship by van to Coffeyville.

I had to leave before that to find a house to rent, which wasn't a problem in Coffeyville. But before that happened, when Helen was only six weeks old, there was a knock on the door one Saturday morning. When I opened it, I was nearly knocked down by Nona, LaRue's mother!

She kept saying over and over, "Oh! I came to see my granddaughter!" No, "Hello, Joe" or "Go to hell!" — Nothing! She just ignored me. The cab driver dropped her suitcase at the door and left.

Nona had taken a train from Corsicana to San Antonio. She stayed a few days, totally ignoring me! That same Saturday that she arrived, we had a stag party at the Wolfe Inn on the old Fredericksburg Road, complete with drinks and T-bone steaks and a floorshow, all for two bucks each!

I came home, let myself in the door, and tripped on a scatter rug and hit both sides of the doorway. I never drank very much but I knew if we ever went to Corsicana, there would be a few women in Nona's Sunday school class who would be informed of my "shortcomings".

They were an elderly group and well informed about the other people in town. To them, secrets were too good to keep or not worth keeping. A common phrase I remember hearing was, "Please pray for me. I have a terrible burden to carry." That was wonderful because they were all involved and had something to share later in testimonials!

Later, we received a letter from LaRue's father, asking us to come for a visit and telling us that there would not be any problems. I reluctantly agreed because I knew he wanted to see his granddaughter. We began going to Corsicana from San Antonio, a five-hour drive, on major holidays or whenever I had some time off.

We never went empty-handed. We took a lot of food LaRue had prepared ahead of time. I managed to make myself scarce while we were there by going fishing or repairing things outside the house.

Hugh's partner in the cattle business was a real nice gentleman, just like Hugh. He and I had a lot of talks at their business location, "The Wagon Yard", behind the buildings in the center of Corsicana. There were a lot of sheds and space for the cattle and their office was there.

Hugh and I became good friends and one day he said, "Walter Beaton said he really likes you and that you aren't like the bad guy he heard people talking about!"

"Nope, Hugh," I replied, "I haven't started growing horns or a forked tail yet!"

I brought LaRue and Helen to Corsicana and left them there while I drove to Kansas to my new assignment. LaRue was expecting another baby and couldn't travel that far.

When Ellen was a month old, I drove down to Corsicana to get them. They loved the large, nice home. There was plenty of space to play and have fun.

When I reported to the Base Commander's Office, a Major, Jim Green, was the Base Commander's Adjutant. I walked in and saluted. "Warrant Officer Joseph Diviak reporting for duty, sir!"

Major Green returned the salute and said, "Who the hell sent for you?"

That burned me up so I dropped my bombshell. "Col. Johnson, the Inspector General's Office, Randolph Field, at the request of Colonel Buck, your Director of Training, sir! I do not appreciate this type of reception, sir!"

Major Green's mouth flew open, but no words came out. At that point, Colonel Harmon, the Base Commander, who was in the next office, came out and put out his hand and called me into his office. Major Green was still looking stupidly around the room!

"Sit down, Mr. Diviak," Col. Harmon said. "I'm glad you're with us. Welcome aboard!" As I sat, he continued, "We have a Lt. Hale presently in charge of the Instrument Training School but he's being moved to a fighter training squadron. He doesn't know enough about the mechanics of these trainers since he's never had that training. I want you to go to Col. Buck's office on the flight line and meet him since he'll be your boss. I talked to Randolph and you are highly recommended. I have also seen your 201 file with some impressive commendations. Any questions?"

"No, sir," I replied. "It's a pleasure to be here and I'll do my best, sir!"

I saluted and walked out. As I was closing the outer door, I heard a voice call, "Major Green, come in here!" I couldn't help but chuckle and I felt good! I have no respect for people like Major Green.

I drove over to the flight line office to see Lt. Col. James R. Buck. He greeted me warmly and told me that the last general inspection of the Instrument Flying Trainers Department by the Inspector General from Randolph was very bad. He had asked the Personnel Department at Randolph to send a trained officer to run the Trainer School. Col. Buck didn't know that the Director of Personnel was my old boss, Major Bailey—now Lt. Col. Bailey!

From there I went over to meet the personnel and inspect the trainers that I would be in charge of for the next couple of years, or maybe less. There were ninety instructors, two clerks, and one custodian. There were thirty-four trainer simulators, with seven out for repairs, which was inexcusable!

I looked over the personnel. Some lacked sufficient training in certain areas. I immediately knew I had a problem here. The atmosphere was far too relaxed and there was poor coordination between the six air conditioned buildings.

I left and went back to my room at Sgt. Hogan's father's boardinghouse. I got out some paper and went to work. I had a Sgt. Tailor and a Sgt. Johnson as clerks and I started with them.

My office was a room in one of the buildings with no doors, only open doorways, and I wanted it that way. I decided to put the Flying Cadets' files and records in four-drawer file cabinets outside of my office. One drawer was for officers. All cadets coming in would be checked in and assigned a trainer and building number.

I put out a written order that Sgt. Hogan would be Chief of Maintenance. I had a Tech Sergeant in charge of all the instructors but he couldn't handle people and was again untrained. I looked through the personnel but there wasn't a leader in the whole bunch!

One day, a Staff Sergeant, Roy T. Thomas, requested permission to see me. He was working on the flight line but was well qualified to be an instructor in our Department. I called Major Gregory in Personnel and had him transferred in to work with me. Major Gregory turned out to be one of my best friends there.

Sgt. Thomas was a wonderful instructor and the best maintenance man on the trainers. One Saturday, he and Sgt. Hogan and I worked all day and had all 34 trainers flying by the end of the day! I kept a pair of enlisted man's coveralls in a locker in my office just for that purpose.

Within about three months, we were in excellent shape. Colonel Buck, our Director, and the Commander were both very pleased.

At 3:00 P.M. every Tuesday and Thursday, we were required to go to Physical Ed classes, which was no problem. We played baseball, tennis, badminton, or basketball. I saw Major Green there one day. He said, "Mr. Diviak, you almost got me fired when you reported in!"

"No, Major," I replied, "It was your rude and crude manner of welcoming me that was your problem! And judging by your remark to me just now, you haven't learned from it, sir!"

"You must be a spy for the IG from Randolph, Diviak!"

"If that is so, you'd better treat me with more respect! Colonel Buck had me promise to come to him if I have any problems with anyone here!" I replied. "Have you ever been to Alaska, Major?"

I never had any more problems with Major Green but later, by accident, he was sent to Alaska! That must have caused him a lot of pain since he hated the snow and ice of Kansas! I often wondered if he blamed me for his transfer!

We had to keep in shape and also had to qualify on the firing range with a Colt .45 automatic, a .38 Special, a Carbine, a Thompson machine gun, and the "grease gun", as it was called—a cheap variety of an automatic gun with thirty-round clips. I qualified with all of them. We also had gas mask drills at any time during the day or night.

The course I liked best was "dirty fighting", closely related to judo and karate. The instructor taught all the kicks to the groin, breaking eardrums, chops to the Adam's apple, and how to crush an opponent's instep with a stomp. He saved one just for himself "in case he encountered one of us in civilian life" he'd have one move we didn't know!

Another course was how to take a gun away from an opponent, even if he was behind you and had the gun in your back. They also taught how to use finger jabs to the eyes and how to use a two-foot long by two-inch diameter club to disable an opponent in half a dozen ways.

We saw a lot of training films on self-defense and how to be the aggressor. Most of all, we were taught not to just charge into a place and tear down a picture or pick up war booty, since any of these could trigger a booby trap. We also saw movies showing the results of such foolishness.

One sergeant, in combat, showed his squad how to make a hook out of wire and a long piece of string to hook a German Luger pistol lying on the ground. After he hooked the pistol, he jumped into a foxhole and pulled the string back. The pit was booby-trapped but not the pistol!

We were all subject to be called to the front. Even though we were in the Air Corps, we still belonged to the Army.

We had weekends off and I had a chance to take my family to the parks for picnics, use the swings at the school playgrounds, and take walks out in the countryside.

Sgt. Charles L. Hogan (we called him "Pat") came from San Antonio with his wife, Ruthie, and had me come to his stepfather's ranch, owned by three brothers, Lawrence, Clarence, and Homer Lowry. The ranch covered over 30,000 acres with a huge ranch house built on a hilltop near a paved road. The house had a full basement, called the "bunk room" which had ten bunks and a huge gas log fireplace. There was a gas well on the ranch.

There were thirteen large stock ponds or lakes on the ranch, all loaded with fish. The Verdigis River formed the boundary of the ranch. Old Grandpa Lowry had claimed and bought the land after the "land boom", while it was still Indian Territory. Knee-deep grass covered the whole place and there was a creek running through the middle of the ranch. It was a fantastic place. A lot of cattle from the northern ranches wintered there.

It was also great fishing! On Saturdays, we refilled our refrigerators with fish fillets, quail, doves, and swamp rabbits, which made the best stews.

Sgt. Hogan's father, Charles, was a railroad telegraph operator at the station in Coffeyville. Charles and his wife, Ellen, owned a large house with apartments on the second and third floors. I boarded with them for a month while I looked for a house to rent. Even though I was a guest in their home, I insisted on paying board because I knew the Hogans would be retiring in a few months and would need all the money they could get. They and Grandpa Hogan, who lived with them, were all very nice friendly people.

Once a year, there was a cookout at one of their relative's homes in the country and my family was invited. I marveled at how nice it was that former husbands and wives could attend these gatherings without any bitterness. They were all great friends, without living in the past. And they were a fun bunch of people.

My father was still a coal loader in West Virginia, but he could take a couple of weeks off if he wished. I had never heard of anything like that before! This was in 1944. He wrote me a letter in his broken English—he would like to come to Kansas! I immediately wrote back, telling him to come on and just let me know which train he would be on.

I picked him up at the Coffeyville Railroad Station and took him back to our house, about half a mile from the station. Dad was so happy to see our little girls and he had brought gifts for them.

We showed him around the wheat fields, the 30,000-acre Lowry ranch, and took him hunting and fishing. He really enjoyed all of it.

I also took him to visit my Instrument Flying School, which amazed him. All of the other instructors just had to meet him. Dad had no trouble talking to them in his broken English.

Warrant Officer Joseph T. Diviak and part of his ninety instructors at the Instrument Flying School at Coffeyville Air Base, Kansas during World War II. (1943-44)

Warrant Officer Joseph T. Diviak C.O. of the Instrument Flight Simulator School in Coffeyville Air Force Base, Kansas

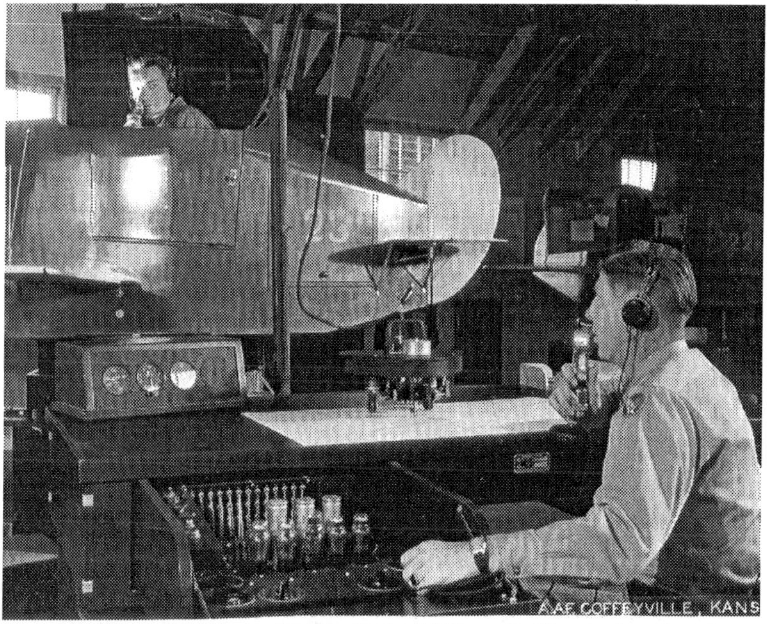

Instructor Earl Libby pre-flights Cadet Wilbur Collins before he closes the flight simulator's hood. The flight recorder on the desk traces the flight pattern on a chart, or map, in ink. The instructor notes on the chart any deviations or errors such as spins, stalls, air speed, altitude, and direction accuracy. There were thirty four flight simulators in six dust-free, air-conditioned buildings.

The married sergeants and their wives gave a party at the mansion Sgt. Penn had rented by luck, which had a large room, about 20 feet by 30 feet, which the owner had used for parties before the war. My gang rented a Nickelodeon and we brought lots of steaks to grill, along with a keg of beer, wine, and other spirits.

Dad danced polkas, waltzes, and other dances with the sergeants' wives. They really spoiled him with attention and he loved every minute of it!

Mom was back in Caretta and Dad had let her know he was okay. In a few days, he received a letter from Mom telling him to come home, that she needed him and that he was having too much fun!

When he boarded the train for home, he hugged all of us, and then he said, "I glad you no work coal mines no more. You say it good when you say 'when I get big I leave here'!" He had remembered and was proud.

My brother John was in the Army Artillery at Camp Carson. He was in a mule-packing outfit, preparing for the invasion of Japan.

My brother Steven was in the Navy, a Crew Chief on a PBY (Air Corps B-24), on patrol in the North Atlantic looking for U-boats.

Our baby sister, Josephine, was in the navy—WAVES—working at the Jacksonville Naval Station in Florida.

Our parents had four of us in the service to worry about.

Before my father left to go back home, he asked if I knew about John L. Lewis being fined $500,000 for calling the coalmine strike during my time in the Army Air Corps.

"Sure I did." I replied. "And I sent $2.00 to his Washington office instead of the dollar the Coal Miners Union had suggested each of the coal miners send to help pay his fine. There were over 600,000 miners. He wouldn't have a problem paying the fine."

Dad looked surprised. "You think he did right by calling the strike?"

"Yes. The coal companies had stockpiled millions of tons outside on the ground and just wanted to get rid of Mr. Lewis. It was all politics. President Roosevelt appointed a judge named Goldberg to handle the case and set the fine. The coal miners saved a great and wonderful man from going to prison. The judge misjudged the coal miners!"

Dad laughed and reminded me of how several times, the coal companies had gotten together and offered Mr. Lewis double what he was making to come work for them as an advisor, a counselor, and how he'd refused!

One last episode in training flying cadets and officers to fly instrument: a First Lieutenant David Perlman, who represented the Inspector General from out training command but was stationed in Coffeyville, came to the School. I saw him coming, stepped out of my office, and greeted him in the outer office.

I had an understanding with the two men who worked for me—when I came out of my office to greet someone it was a special case. They knew that I wanted either Sgt. Thomas or Sgt. Hogan to be the instructor and to handle that officer's training session in a special trainer setup.

As Lt. Perlman walked down the boardwalk to Thomas's building, I called Sgt. Thomas and told him to use black ink on the notations of all air spins, etc., on the recorders path on the chart and not to give them to Perlman. I said to tell Perlman we would get them reproduced if he wanted a copy.

After Perlman finished and left, Thomas came to my office with the large flight map covered with dozens of notations. I looked at it. "He can't fly instrument as good as the cadets in mid-course training!"

Sgt. Thomas told me Perlman had spun in twice when his air speed had dropped under 110 miles per hour, he lost headings and wandered all over the chart and never even made it back to the air base!

I filed the chart and waited. Finally, one day, Col. Buck, the Director of Training, called and said he wanted to see me. I knew exactly what it was about and took the Perlman file with me.

Col. Buck greeted me warmly and we had a cup of coffee, then he showed me the report Lt. Perlman had sent to our Headquarters Inspector General's office with a copy to the Training Commander's Headquarters at Randolph Field,

I read the report and started to grin. I must have looked too happy because Col. Buck said, "Diviak, you got him, huh?"

"Yes, sir!" I grinned at him. "In black and white! Here is his flight chart laid down by the recorder from the flight trainer simulator, sir!" I handed him the map. "Sgt. Thomas, my best instructor, was assigned to fly Perlman."

Col. Buck went over the map carefully, inch by inch, and finally, he said, "Lt. Perlman has a big surprise coming when I send this through channels and back to his Commander!" He looked at me. "You were ready for him, Diviak, and I imagine you're ready for anyone else who wants to discredit your organization!"

"Anything new in the Army is always subject to ridicule, sir!" I said. "We started this simulator program in 1940 and I was in the very first class. It's been an uphill battle all the way, sir!"

"I appreciate you and your instructors," Col. Buck replied, "And will make sure the training command is informed. Anything else, Mr. Diviak?"

"Yes, sir. A suggestion." I said. "I think we should have several copies made at the base print shop. I need one for Lt. Perlman's file in case there are any problems, sir."

"Right!" Col. Buck exclaimed. "A damn good idea! I'll have my clerk do that. Mr. Diviak, are all Warrant Officers former Master Sergeants?"

"Yes, sir!" I replied. "The two you have here are, sir!"

"Great!" Col. Buck said. "If you ever have any need for a letter of recommendation, let me know. Keep up the good work!"

I saluted and left, fairly floating down the boardwalk back to my office.

Six months later, Lt. Perlman was ferrying aircraft and was doing okay while he was on stateside duty, but as he was ferrying one to the Hawaiian Islands and was never heard from again.

One unusual thing happened. I had to have a Sergeant Clerk transferred because of friction between him and the flying cadets. This Sergeant was very effeminate and spoke so sweetly! He was transferred to the Chaplain's office, where he was very happy.

Staff Sergeant Bill Ashley came in to interview for the clerk's position. He walked with a limp and used a cane. During the conversation, he told me he had been a tail gunner on a B-26 and had stopped two Japanese Zero bullets in his left hip.

I asked him if he knew or had ever heard of Captain James P. Muri, who flew torpedo-carrying B26s.

"Yes!" He replied. "I was his tail gunner at the Battle of Midway Island when I was wounded!"

I was so surprised and told him that Jim and I were great friends.

He looked surprised to hear that and said, "I'll never forget Capt. Muri! He brought that B-26 home after we put a torpedo in a Japanese Aircraft carrier. The plane had over 500 bullet holes in it and the hydraulic system was shot out! We belly-landed at the Midway Airfield and the entire crew received the Distinguished Service Cross. I got credit for three Zero's and then I was wounded going home to our airbase!"

Staff Sgt. Ashley was from Ashville, NC and we became great friends while he was a clerk in my office. Years later, when I told Jim the story, he said, "That is so unbelievable!"

We had completed our mission at the Coffeyville Airbase and it would be closing after the last class of 450 cadets finished. A couple of my instructors were building contractors in civilian life. I called the group together and told them we had to remove the fake wings from all thirty-four trainers. We also had to remove the electronic equipment from the base of each one. Everything then had to be packed up in wooden crates to be shipped to storage in another state.

The two men who were builders were given the instruction sheet on how to pack the trainers and materials necessary to pack one trainer. They then planned out what they needed to pack all thirty-four, went to the post engineering office and got saws, lumber, nails, packing materials, etc.

I helped them and we got the job done in three weeks, using a sort of assembly line setup, and moving the materials to the trainers instead of moving the trainers. It worked.

Randolph Field, Texas was my next assignment. My orders read in part, take twenty-five of your best instructors for change of station and report to Training Command Headquarters, Randolph Field, Texas. I had Sgt. Hogan at the top of my list but they said they "had other plans" for him.

At Randolph, we were to start a school for flying officers who flew transport planes for the Army Air Corps. We would have ten trainers in Hanger #5, which was air conditioned and dust-free, as was required.

My furniture was crated and moved once again to storage. We all said goodbye. I drove my family to my in-law's home, where they were waiting to spoil their grandchildren. Then I drove on to Randolph and reported in. Col. Johnson, my new C.O., greeted me. He was glad to see us because they wanted to get the program under way quickly.

Once again, I went house hunting, this time in New Braunfels, Texas, and found a duplex about ten miles from the base (which is about twenty-five miles from the center of San Antonio. A man owned the duplex who lived next door and who also owned a furniture store on the same grounds.

The furniture arrived and I went to pick up my family, some ninety miles away. With the help of the landlord, Henry Silver, and his handyman, we moved in and were made to feel comfortable.

New Braunfels was a small town of about 3,000 back then. It had originally been settled by German immigrants in the early 1800's. The Duke of Braunfels brought three hundred families over on a ship that landed at Galveston. From there, they rode in wagons drawn by horses to settle the area. The rulers in Mexico had given them Spanish land grants. Even now, a lot of the property in that town can be traced back to those land grants in the 18th century.

New Braunfels was and still is a beautiful place. There is a big spring flowing where a river starts, the Comal River, in Landa Park. Several million gallons of cool clear water flow out of a hole at the base of a hill every twenty-four hours. Farther downstream a flourmill with great water wheels used this water to turn the grindstones.

We fit in very nicely with the people there and they were very nice to us.

We flew the trainers from 3:00 P.M. to 11:00 P.M. because there was less noise and traffic on the base then. I liked working in the evenings.

LaRue and our daughters, Helen and Ellen, and I would take a small folding table and chairs, a charcoal grill and food and have our breakfast at eight o'clock in the morning on the pretty grass banks of the river under the huge live oak trees. We all loved it there and generally had the whole area to ourselves and it was only a mile from our apartment and easy to get to.

We fished the Comal for Rio Grande perch and bass each morning and had a fish fry a couple times a week. We often invited a couple from the other side of the duplex to share it with us.

Henry Silver was always buying things in “lots”, like three or four bushel baskets full of radio tubes, damaged toasters, roasters, and such. I took the tubes to our back porch one day and used my volt-ohmmeter to check the filaments. Over ninety percent of them were fine. Henry got them for fifty cents a bushel and sold the tubes for fifty cents each!

He seemed to make money on whatever he bought and sold. I repaired a lot of the electric appliances for him and he would bring gifts over for LaRue and the girls.

Six months later, I received orders that we had completed the project at Randolph and I was to take the twenty-five instructors and go to Bryan Air Base in Texas. We were to help train more instrument flying instructors there and the flying officers would then train other officers in advanced instrument flying.

All the instructors received their orders and traveled by bus or private vehicles to get there. I reported in after spending the night at a hotel. I was warmly received but, strangely, I felt that I wasn’t needed. They had more officers there than they needed.

Major Foley from Napoleonville, Louisiana was our Squadron Commander and Col. Duckworth was the Base Commander. The place was built on land that had once been a cotton field and was only three miles from the city of Bryan. The only quarters for rent were in Caldwell, about eighteen miles from Bryan Air Base. Caldwell was a Polish, Slovak, Moravian, Czech, and German town. Noblemen who brought three hundred families at a time from Europe settled the entire area. Each family was given land by the Mexican government in the early days of the settlement of Texas.

My Chief Instructor, Staff Sergeant Roy Thomas, found a house half a block away from ours. We settled in and were kept very busy at the airbase. Most of my time was spent signing papers and keeping records and taking care of schedules. There were only half the trainers that I'd had at Kelly and Coffeyville.

We had Phys Ed every Thursday, rifle range once a month, and I drew a lot of Officer of the Day duties. I knew it wouldn't be long before this base was closed, too. I had an additional duty as Supply Officer and made inspections of the barracks and trainers.

The people in town were nice and they all appreciated our business because there weren't any factories there. The primary income was cattle, cotton, and farm produce. College Station was six miles away, where Texas A & M was located. I enjoyed a few football games there. We hunted ducks on the ponds near there and I did a lot of fishing in the two rivers near Bryan.

The Brazos River was big and dangerous. We took great care not to take too many chances "bank fishing" there. The Navasota River that joined the Brazos about twenty miles south of Bryan was much better. The catfish were delicious!

In three months, I found a large house for rent that had a "mother-in-law" three-room apartment upstairs. I rented the house and sublet the upstairs apartment to Sgt. Thomas and his wife, which worked out fine for all of us.

I drove and Sgt. Thomas, Lt. Phil Jensen, and Lt. John Kloate, who all worked in the same department, rode along with me. I had no trouble getting gas ration stamps. You had to have them to buy gasoline.

It was getting near the end of the war and for some strange reason, a lot of schools were kept open, like San Antonio Air Command (SAC) for Warrant Officers and Commissioned Officers to attend six weeks of Squadron/Company Commanders School at what is now Lackland Air Base, adjacent to Kelly Field and Media Nuclear Storage Facility in San Antonio.

In 1944, I was ordered to attend the school. I caught a train to San Antonio and was picked up at the station by a truck form SAC.

I reported in and was assigned to a barracks. There were eighty of us. We learned to drill with rifles, take charge as squadron or company commanders at drills in marching and in review. We learned map reading, hand-to-hand combat, and got lots of exercise every day. Everywhere we went it was always in formation.

The climax was an infiltration course, at night, through barbed wire barricades with trip wires to explosives. I was glad to get out of there because of the rattlesnakes and the explosives in the "pits" near the barbed wire. It was rugged training and so unnecessary because there were surplus officers at every base!

Through the years, I had stayed in touch with Paul and Muriel Boatman. He was the Chief of the detachment of Department of Commerce at Tucson while I was there. He was later transferred to Burbank Airport in California, then to be Chief of the Fort Worth Region in Texas.

He told me that they were planning a Federal project to install glide slope landing equipment throughout the four-state southwest region. It intrigued me so I applied. I was well qualified for the job.

When I returned to Bryan, LaRue and I talked about it and she said that she was tired of moving. We knew that I would be sent overseas as a member of the Army of Occupation. The war was coming to an end. I had corresponded with friends in Europe and they said all they did was maintain order and patrol the red light districts to make sure that American troops didn't visit those places. It was a boring job.

CHAPTER SEVENTEEN

A Civilian Again

I had acquired enough points to get out of the service, so I did. One December 7, 1945, I was placed on inactive status. I received a half page of commendations for my work at Bryan Air Base, while the man who wrote mine for our Commander wrote himself three pages of commendations! It was unbelievable but typical in those days in the service. I had a large file at Personnel.

I drove to Dodd Field in San Antonio and went through the Processing Center, which was mostly medical. At the end, I was handed a Certificate of Service and a lapel medal, which we called a "ruptured duck". It must have been meant to look like an eagle.

I returned to Bryan and put our furniture in storage until I secured the job in Fort Worth. Two weeks later, President Truman cut the budget and the Glide Slope Project was cancelled. I was suddenly out job hunting, in Corsicana, Waco, and Austin, but there were no jobs available.

We were staying with LaRue's family in Corsicana and helped buy the groceries. I did repair work on the house in between job-hunting trips.

One Thursday evening, I received a call from my old friend and buddy, Staff Sgt. Roy Thomas. He had also left the service and had taken a job as an electrician with C.E. Grieser Electric Company in Bryan. Roy told me that he had spoken to Charlie, the owner, about me looking for a job. Charlie had told him to call me and have me come down to Bryan, some ninety miles away, for a job interview on Saturday morning at 10:00.

It isn't biblical, but whoever said, "If you cast your bread on the water, it shall return sevenfold" was right!!

I remembered Roy coming to me looking for a job in the Coffeyville Instrument Flying School. I interviewed him and accepted him immediately when I learned that he was a graduate from the Instructor's Trainer School. When I left Kansas, I took him with me to Randolph Field, then to Bryan Air Base, and found housing for him and his wife, Katherine.

I drove to Bryan for the interview. As we walked through the workbench area in the Electric Company building, I noticed shelves and benches covered with appliances needing repair. "Do you know how to repair them?" Charlie asked.

"No problem." I replied. "And I'd like to start Monday!"

He laughed. "You have a job, Joe. Welcome aboard!" We shook hands.

I rented a room and worked a week then visited my family in Corsicana and picked up more clothes and returned to Bryan. I found a new little two-bedroom house with a garage and a nice back yard. I qualified under the G.I. Bill of Rights to purchase the house. We bought it and moved in.

We fenced the back yard and had a garden and a small chicken yard, since we were at the edge of town. We bought ten chicks, but in a few weeks, they were large and turned out to be nine hens and a rooster. We had all the eggs we needed and even sold eggs to the neighbors, which easily paid for the chicken feed.

One day I was working in the back yard and noticed that one of the hens had her whole right side ripped open just behind the wing. I went into the house and got a bottle of merthiolate, some cotton swabs, a large needle and some regular white twine, the kind used by the grocers to tie up meat purchases—no Scotch tape in those days!

LaRue and the children went for a walk and I caught the hen and sat down on the back steps. I began sewing up the long "L" shaped gash, pulling the skin together, and making certain no feathers got into the wound. It was a painful process. The poor hen squawked every time I took a roll stitch but didn't struggle too much. I had her head under my left elbow.

Then I looked at the rooster's spurs, which are used for mating and fighting and knew he was the culprit. Some of our neighbors had complained that he crowed too early and too often. He ended up in a large pot with dumplings, which we shared with our neighbors. That solved the problem.

The wounded hen was back to laying eggs within a week and sure earned her keep!

I was going to school two evenings a week at the Electric Company under the G.I. Bill, learning to be a Journeyman Electrician. Being a radioman, it was very simple for me. Charlie Griesser, who owned The C.E. Griesser Electric Company, had a Master Electrician's license. The company was not only a repair shop for appliances; they also did rewiring of old homes, commercial buildings and new homes as well as factories and service stations.

It was a prosperous business and with a Frigidaire franchise dealership, did quite well. Mr. Charlie was a hands-on boss. He was the honorary Fire Marshal for the city and oversaw the training of the fire station personnel and updating the equipment. He was given a bright red Oldsmobile, properly marked and with a flashing red light, to attend fire emergencies.

I never met a man who was more honest than Charlie Griesser. His handshake was his bond. We had about nine employees most of the four years I worked there. We were always busy because "Mr. Charlie" was well known for his honesty and integrity.

Bryan, at that time in 1946, was a small city of 20,000 people. Roland Dansby was the mayor. Bryan Air Base had been there during the war. I had been stationed there as Link Training Officer and school secretary for training instructors. When the base closed, the property became part of Texas A & M University's 3,600-acre campus, the largest in the USA.

Bryan was a fun city. We all enjoyed the cookouts and get-togethers at the electrical shop on holidays and the other family-oriented activities. We men were all fishermen and helped the two game clubs catch enough fish to feed the approximately 150 people at the fish fries and barbeques. Everyone worked together and stayed in touch.

We were members of the First Baptist Church and I went to the Businessmen's Bible Classes, which had 150-160 members. We met in a large room in the basement of the church and generally, there were between 85 and 90 members there every Sunday morning.

Dr. P.G. Gunter, a professor of English at Texas A&M, was our Bible teacher. The classes were composed of businessmen, teachers, lawyers, laborers, etc., and a lot of them were from different faiths and different churches near our church. Dr. Gunter was the best at getting the attention of the class and making both the Old and New Testaments live again as if they had happened yesterday.

We managed to take ten days off and make a trip back to my home in West Virginia to see my parents, who had retired to a home in War. When you retired, you had to move out of the coal mining towns.

It was a two-day drive from Bryan to Caretta and War. All the roads went through the center of most cities and we would drive nine or ten hours a day. There were nice motels and good restaurants along the way.

My parents were overjoyed to see and hug their granddaughters and to meet LaRue. We had a great time together.

Mr. And Mrs. Joseph E. Diviak retired in the 1950s after he worked in the coal mines for forty-four years as a coal loader. They reared nine children and enjoyed many grandchildren in their retirement in War, West Virginia.

After moving to Bryan, Helen, our oldest daughter, then three years old, developed a spotted skin rash on her forearm that just would not respond to treatment. LaRue took her to a doctor who had a great reputation as a healer. His office was near the Electric Company. LaRue called me at work and said that Dr. Brenner and his assistant, Dr. McGill, wanted to talk to us immediately.

We met in his office and Dr. McGill said, "We have some real bad news for you, so please be seated." After we sat, he said, "Helen has acute lymphatic leukemia, which is an incurable disease."

We sat there in shock. I asked about treatment and he told us that it could go into remission but that would not be permanent. Then he gave us a letter and sent us to Dr. Longmire, a diagnostician at Scott and White Hospital in Temple, Texas. He confirmed that it was leukemia and gave us instructions on how to take care of her and prescriptions for medication to give her to make her more comfortable. The prognosis was about two years.

Helen lived two years. She had seven blood transfusions. All the blood was donated by volunteers–cadets at Texas A&M. Only seven people of the 2,500 students there had Helen's AB negative blood type. All seven volunteered but only two were used. None of them would give us their names so we thanked them through the doctors. I will never forget the "aggies" and the mere mention of Texas A&M still generates a warm feeling in my heart.

The last two years were rough for Helen, who was a blonde, blue-eyed doll of a little girl. She died just two years after the first diagnosis and is buried in Bryan.

The people in Bryan were so generous in their assistance and constantly wanted to help. The Sisters at St. Joseph Hospital took turns with us at night when Helen was there and did their best to help us, but most of all, Helen.

I returned to work the day after the funeral. Mr. Charlie told me to wait a few more days, but I knew that the best cure was work and so did LaRue. Ellen was two years younger than Helen and she was a great help. She seemed to know that we had all we could handle and never gave us any problems. She stayed busy playing with her toys and dolls.

Someone in town kept sending us a certified cashiers check to help pay expenses. It was a great help to us. We had life insurance, which paid all of the large hospital bills after Helen died.

"Mr. Charlie" let me buy a 16-inch oscillating fan at cost and we gave it to the Sisters at St. Joseph's to use in their sleeping quarters, which were behind the hospital. It did get hot in Bryan. They all thanked us many times. It was really needed. There was no central air conditioning in those days.

I worked in the electrical shop repairing anything electrical—tools, appliances, radios, sound equipment, incubators, iron lungs, and so on. The farmers loved our shop. The area around Bryan was farmland, growing mostly cotton, corn, cane, and maize along with raising horses, cattle, and hogs.

Over the years, the land changed hands in some areas and yet some managed to hang on to the prime land on the Brazos River Plateau, which is a vast network of cotton farms today.

The George Chance family was a great example of keeping the large acreage in the family. George had a brother who was born mute but had still managed to secure an education. Clarence always wanted to be called "Dummy" Chance and often came in the electrical shop for parts and electric cables for the farm on the Brazos River. We became friends.

He could read lips but would always write a note in reply on a small pad he always carried. One Monday morning, we had heard about a poker game over a weekend at a ranch house where one man had lost a bakery, another a café, another a service station, and so on. He came in and I asked him what happened to the families that had settled there on the land grants.

Clarence laughed and wrote: "The family trees have died and left the saps here!"

My boss had me attend Texas A&M University's Monday Morning Quarterback Breakfast, sponsored by the Alumni, Boosters, and businessmen in Bryan and College Station.

I went in his place most of the time. It was fun to hear Coach Homer Norton go over the film of the previous Saturday's football games A&M had played. I love it and learned a lot about football there and met a lot of nice people.

About three years after Helen's death, Barbara Jean was born and she was a live wire from the first day! She had her own agenda and had a lot of things she wanted to do! At age five, she learned to roller skate and to ride a bike by herself. Did not need any help! She brought a lot of joy to our family and Ellen greeted her with open arms. We still missed Helen.

Roy Thomas and I fished a lot. The Navasota River flowed east of Bryan and the Brazos west of the city. They met about twenty miles south of Bryan. It was great fishing for perch, catfish, and freshwater drum. We kept our families well supplied with fish.

We needed a light boat for the Navasota because the banks were too boggy and not very accessible. We built two galvanized steel boats, 45 inches wide, 14-feet long and 18 inches deep, which took a lot of crimping and rolling the metal seams. But trimmed in oak wood, seats, and five-horsepower motors, they were perfect for what we needed.

My father-in-law found a beat up two-wheeled cattle trailer for me to use as a boat trailer. I converted it with lower sides and used it not only as a boat trailer, but a camper and to move our canned goods, tools, etc. when we moved from Bryan. The moving company took the furniture.

"Mr. Charlie", who was also the Fire Marshal, was listening to the local radio station announcer who was doing a commercial for the electric shop. The announcer, Bill Zak, was also a comic and said: "Take your electrical appliances, ladies, to the C.E. Griesser Electric Shop and give them to Joe Diviak. He can repair anything! Don't give them to Charlie, though—all he can do is put your fire out!"

That commercial really upset Mr. Charlie but after all the kidding, he got more publicity than he ever dreamed of for our repair business. We were swamped!

Mr. Charlie's mother had a bakery in the back of their huge home. She and Mrs. Griesser ran it with help. As one walked into the electric shop, to the left were showcases filled with pastries, cakes, bread, and such. To the right as you came in the door were Frigidaire refrigerators, stoves, dryers, and washers plus all sorts of small electrical appliances.

A door in the back of the bakery opened into the repair shop and a storage area for wire, conduit, and other supplies for building electrical supplies, plus refrigerator and deep freeze repair supplies. A hand-and-rope operated elevator went to the second floor and there was a staircase as well.

Double back doors opened from the shop into the alley and a large gravel lot. Most people came into the shop by the back door because of the easy access from the parking lot with their appliances.

Mr. Charlie had two daughters and a son. It was remarkable how nicely they all lived together in their large home, with two mothers-in-law sharing a bedroom! They were really sweet to each other all the time and I knew them for four and a half years.

Bob, the oldest child, went to Texas A&M and he loved gadgets. He was always experimenting with electrical gadgets when he helped in the shop on Saturday mornings.

One day, while his father was away on a trip to Austin to attend a Fire Marshal's meeting, Bob filled a huge weather balloon with methane in the back of the shop. He took a ten-foot length of light, half-inch conduit and taped a lighter on the end of it. Then he attached a string to the balloon and let it float upward fourteen or fifteen feet, then touched it with the lighter flame.

There was a tremendous explosion heard all around the center of town. People on the streets ran into buildings for cover. People inside the buildings ran outside to see what had happened. No one was hurt but Bob was a bit shaken up and stunned for a while!

Of course, Mrs. Roundtree, the editor of The Bryan Daily Eagle, the local paper, ran an appropriate article, which stated: "While Mr. C.E. Griesser, our fire marshal, was in Austin attending a state fire prevention meeting, his son, Bob, was at the electrical shop and store, experimenting with a gas-filled balloon and a lighter. He darn near burned the town down!"

Bob was grounded for quite a while and calmed down quite a bit after that!

Mr. and Mrs. Roundtree were the owners of The Bryan Daily Eagle and after Mr. Roundtree died, she took over as editor. Everyone loved her! She was always so sweet and nice to everyone she met in her office, the restaurants, or on the streets.

She was also a great member of the large Lutheran Church. The pastor, Reverend Braunfels, had done a great job of enlarging the church with new additions to the building as well as increasing the membership in the four years he was pastor there.

Finally Rev. Braunfels received orders from the Bishop that he would be moved to another church, as was the Church's custom to move a pastor every four years. This created a huge uproar and a lot of ill feelings among the congregation.

Mrs. Roundtree rose to the occasion in a wonderful editorial: "We will all miss Rev. Braunfels, who has been like a father to most of us. He has done a lot for the church, the county, and the City of Bryan. We will all be sad to see him leave and will sorely miss Rev. Braunfels. But we should not be selfish and not share him with other churches that need him.

"Of course, when we receive a pastor who is not so good, we can also share him after four years, so the system has its advantages!"

Everything became peaceful after that and the new pastor did a great job too and they were all happy with him.

Roland Dansby, the owner of a large steam laundry, came into the shop with some electric steam irons he wanted repaired. In the course of conversation, I asked him why the city wasn't growing. I knew the closing of the air base put a lot of people out of work, but it seemed to me the town should be growing. His reply was a classic: "Joe, we need a few first class funerals in this town. We have people with money who want to keep it this way!"

A few years later, Roland became the mayor of Bryan and the town was growing in all directions. I was in town on a visit and ran into him on the street. After we shook hands, I asked, "What happened?"

"We had those first class funerals," He replied, "And it is wonderful!"

We had a young apprentice who wanted to learn the electrical trade. Bill was full of gossip about a lot of the leading citizens in town that no one dared to make public. We had an elderly gentleman who rewound electric motors that had the coils burned out. One day, Mr. Frazier was working on a motor on the long workbench near me when Bill came in beaming.

"Mr. Frazier, did you hear the latest?"

"Look, Bill," Mr. Frazier replied, "I spend seventy-five percent of my time minding my own business and twenty-five percent leaving other people alone. I do have a great admiration for people who do the same!"

Bill got the message and a few weeks later, he and his parents moved out of town.

The city of Bryan was and is one of the nicest cities I've ever lived in. The people are unbelievably friendly and get together socially more than in any other place I ever had the pleasure to live in all my years of moving some fifty or more times.

Capt. Malcolm Wall and his wife, Opal, were great friends of ours during my Army Air Corps service at Bryan Air Base. He was later transferred to work on a special project involving long, high-powered radio wave transmission that was being developed at Point Barrows, Alaska.

Capt. Wall stayed there for two years after the war before returning to Bryan and his family. He was placed on inactive status in the reserves. His civilian job was with the Western Union office in Bryan, which he didn't like, but jobs were scarce after the war, so he took what he could find.

LaRue and I invited the Walls to dinner often at our house, which seemed to help Malcolm to adjust. He seemed to be in a daze from the loneliness of those two years in Alaska but became more open and happier after each visit to our house. This made Opal feel that she was indebted to us, she was so thankful.

Malcolm missed the Air Corps and after a year, he re-enlisted and was sent to an air base in San Antonio. Opal and their daughter Laura moved there too. Opal was a registered nurse and she soon got a job at one of the hospitals in San Antonio, where she met a lot of influential people.

I had worked for the Electric Company for four years and was eligible to take the Journeyman Electrician's exam. This meant I would have a license that would require Mr. Charlie to pay me a better wage.

I talked to Mr. Charlie about it. "Joe," He said, "Our Master Electrician has been with the company for twenty years and is in great health. You are now earning just a few dollars less than he is, and I cannot afford to pay you a Journeyman's salary."

Then the most amazing thing happened, as always.

That evening, I received a call from Opal. "There was a notice in the San Antonio newspaper," She said, "that said: 'Anyone wanting to take the exam for an electrician's license, please call the San Antonio Board of Examiners for an appointment to be tested for qualifications on Tuesday of the following week at 8:00 A.M. The applicant should bring a letter of recommendation and verification of his four-year apprenticeship, and be prepared to answer personal questions, etc.'. I called and entered your name!" She said.

I knew I was being offered a break, maybe my only chance to get out of a "dead end job" as they were called in those days. There were certainly no promotions in sight!

I told Mr. Charlie about it the next day and he was really pleased. He had the secretary/bookkeeper type up a wonderful letter of recommendation along with a statement that I had finished my four-year apprenticeship as an electrician.

I put my resume together along with other things I thought I might need for the interview to qualify to take the exam. On Monday afternoon, I drove to San Antonio and checked in to a motel about two miles from the office. Tuesday morning at breakfast, I reviewed my qualifications.

I was at the Chief Electrician's office before 8:00 A.M., checked in with the secretary, and waited to be called. Finally, I was shown into a conference room where the Qualifying Committee of five men was seated around a conference table. I introduced myself and each of the men introduced himself to me.

They took my resume and the letter of recommendation, and then asked me why I wanted to move to San Antonio. I explained the position I was in at my job, that there was no chance of promotion. Then they asked a few questions about my electrical skills.

After about twenty minutes, they asked me to leave the room while they had a conference. About ten minutes later, I was taken back to the conference room. The City's chief spokesman of the Review Board told me that I was well qualified to take the exam. He told me to be there the next Wednesday at 8:00 A.M. and to bring a sandwich—it would be an all day affair. Once the exam started, I would not be permitted to leave the room except to go to the restroom or the Coke machine out in the hall.

Before I left San Antonio, I bought a book on the Simplification of the National Electrical Code, which explained the laws governing electrical wiring in more detail. I returned to Bryan a very happy man.

The following Wednesday I arrived at the examination room fifteen minutes early. There were several other men there to take the exam—one for the third time and two for their second try.

Vacant tablet-arm chairs separated us and each of us was given a large booklet. After we had filled in our names and such, the exam began.

The exam was a breeze for me until I got to a question asking for "the smallest gauge wire for festoon lighting". It sounded to me like a string of lights strung up without a cable or other support, such as those for festival lighting strung across a street. That would be 12-gauge wire. There were all sorts of electrical drawings to complete as well.

I finished before noon and checked all my answers again before I turned the exam booklet in.

Mr. Mahan, the city's Chief Electrician, looked surprised when I handed him the booklet. He asked if I was having a problem and I said no, that I had finished the exam and had checked all my answers several times and I was finished.

He still looked puzzled and scanned the booklet. Then he told me to call his office on Friday for the results. We shook hands and I left for Bryan.

LaRue, my very supportive wife, looked surprised to see me home so soon, but she knew I must have known all the answers to be home so early.

On Friday, I called Mr. Mahan. "Joe!" He said, "You 'aced' the test! Congratulations! Send us $5 and we'll send you your Journeyman Electrician's License!" Which I promptly did!

That evening, I called Opal and gave her the news. She was so happy and of course, I couldn't thank her enough.

A week later, Opal called me and asked me to meet her in front of the Alamo on Friday at 11:00 A.M., and that she would have a family friend, Lt. Col. Jacks, with her. I met them and it was quite a celebration. Opal had to leave to go on duty at the hospital. Col. Jacks and I went to the Civil Service offices on the fifth floor of the post office building.

We entered an outer office and Col. Jacks told the thin, pale-faced clerk sitting there that he wanted to see Mr. Stein. The clerk stared at him, then asked in a squeaky voice, "Do you have an appointment?" No "sir" was used at the end of his question.

"I don't need an appointment!" Col. Jacks replied and we walked past the protesting clerk and into Mr. Stein's office.

"David, I want you to meet Joe Diviak. Joe, this is Mr. Stein!"

"Nice to meet you, Joe and it's great to see you again, Earl! What can I do for you gentlemen?"

"Joe is an ex-Army Air Corp radioman. He served nine years." Col. Jacks replied. "He would like to secure a job in radio or radar in civil service. Do you have any openings?"

"Let me look!" Mr. Stein said. After a few minutes, he said, "No, we don't have any openings like that but we do have a position open for a Plant Journeyman Electrician. Does he have any proof or letters of certification?"

"He has a City of San Antonio Journeyman Electrician License. Would that do?" Col. Jacks asked. "He also has a career resume."

"That is super, Earl and Joe!" Mr. Stein replied. "Joe, I'll give you a letter and you can go out to Kelly Field and meet the Personnel Chief. He'll tell you what to do!"

Mr. Stein and Col. Jacks talked for a while then we left with the letter and headed for Kelly Field, where the clerk greeted us at the door. After we met the Chief, a Master Electrician in charge of electrical maintenance at Kelly Field, met with us, asked a few questions, and Okayed my employment pending a week for a background check.

The following week, I returned and took a physical, aptitude tests, comprehension test, a supervisor's battery test, etc.

I returned to Bryan and gave Mr. Charlie two weeks notice, which took a load off his mind, He told me that his whole family was pulling for me to succeed and that they didn't think there was much chance of Bryan growing for several years.

CHAPTER EIGHTEEN

Federal Civil Service

I returned to San Antonio and rented a small duplex in Highland Hills. We moved there in September of 1950. The duplex was near Brooks Field and Kelly Field was only ten miles away, across the south end of San Antonio.

We lived in the duplex for a while, and then I found a new house to rent within walking distance of Kelly Field. It was a nice house with a chain link fence all around the yard and a two-car garage. It was a short walk to the security gate at Kelly and the hangar we were using for shops and supply rooms.

We all wore ID badges and were physically searched periodically. Lunch kits, thermos bottles, paper sacks, and purses were all looked into as we went out through the security gate manned by civil service police.

I'll never forget my first day at Kelly Field. The Chief Electrician had me come to his office and we had a nice talk. Then he told me he had assigned me to a crew wiring up the new outdoor 4360-horsepower engine test cells back of Kelly Field.

He also warned me not to tell anyone that I had some college and a City of San Antonio Journeyman's License because there were some men there who would make it difficult for me.

I thanked him and kept quiet about my qualification and license. I didn't even tell anyone about my radio and TV repair capabilities.

Our lead electrician had a large wiring diagram of the test cells that looked like a road map, but he couldn't read it.

During our lunch break, he handed it to me to look at. It wasn't as complicated as the wiring diagrams for a radio or transmitter. In fact, it was fairly simple.

I quietly explained it to him—all the switches, relays, and pumps for different octane's of gasoline, oil, etc., that went to the test cells. There would be six engines tested either simultaneously or selectively. All would be controlled from soundproof concrete control rooms equipped with control panels and shatterproof viewing windows. Each engine would have a four-blade, 18-foot club propeller.

The one problem I came across was that the drawing was incomplete. Sixty pairs of wires for the controls that went to the test cell operators and the fuel tanks approximately 200 feet away had been left out.

Over the weekend, I drew an addition to the master drawing we were using. Within a few days, we had one of the new test cells up and running, then another and another, until we had four in operation.

The only thing wrong was that we had to meet every morning at the Chief's office to go over the day's assignments. The new supervisor did a stupid thing, after all the help I'd given him! He had gone to Personnel and looked at my personal 201 file, which was fine, but after one of the meeting, he said, "Hey, Diviak, I didn't know you had a City of San Antonio Electrician's License and a couple of years of college!"

My fellow workers looked at me in disbelief. To most of them, it didn't make a bit of difference, but it did to a few. After that, they avoided me and remained mute whenever I asked them a question. I was very careful, checking my work over and over to avoid any criticism.

Col. Jacks, who was a bachelor and the top engineering officer for our two-star commander, would come to our house for dinner, cookouts, and just to visit. My family loved him like an uncle.

One evening, he asked me how things were going at Kelly Field. "Fine!" I replied, "And I owe it all to you, Colonel! I'll never forget it!"

"I have a friend in the electrical department there," He said. "He tells me you're a great electrician. He also told me what happened at one of the pre-work meetings. Don't worry, Joe, I'll take care of that!"

The following Monday morning at the early morning meeting, Col. Jacks walked in. "Good morning, gentlemen!" We all responded and as he walked by, he motioned for me to follow him, which I did. We went outside and stood on the airplane ramp, talking and laughing, then shook hands, as a lot of men stared at us.

From that day on, I never had any problems with the men. I told Col. Jacks about it at dinner one night at the Billy Mitchell Restaurant outside the main gate at Kelly Field. He grinned at me and said, "I knew my visit would do the trick!"

Within a year, we had finished the six test cells and my ears finally quit ringing. I always wore rubber earplugs during work and at the end of the day sometimes they had to be pulled out with long-nose pliers.

In the meantime, I was assigned a helper and we were rewiring the electric in the lighting over the workbenches. They were constantly renovating to accommodate new equipment. One day, I was up on a twelve-foot ladder in the radio repair facility, repairing some four-foot fluorescent light fixtures.

The Chief of Electronics' office was near where I was working. There were over nine hundred radar and radio technicians under his supervision.

I felt the ladder shake and looked down to see the Chief, Tom Enderly, standing there looking up at me. "What are you doing up there, Diviak?" He asked. "Come in my office. I want to talk to you!"

I came down and followed him into his office. Tom poured us each a cup of coffee from the pot sitting on a table behind his desk, then asked me to sit down.

"Diviak, I've been up at Personnel and checked a few 201 files and I liked yours! I notice that you've been here a year. Why are you an electrician when I need a radar and radio repairman in the shops?"

Tom looked a bit puzzled as I told him what had happened at the post office, where I was told the only vacancy was the Plant Electrician's job.

Then he told me to put in for a transfer and that he would get me within a month. It would mean a $900 a year promotion!

When I told Col. Jacks what happened, he was furious because someone at Personnel hadn't told Mr. Stein, the Civil Service Commissioner, that radar and radiomen were needed!

I walked a clearance through at the Electrical Shop, much to the dismay of the Chief Electrician, but he was glad for me. They had lost the wiring diagram and asked if I would do another. I did it at home and gave it to him a week later.

When I showed up at Tom's office on a Monday morning, he greeted me with a smile and a warm handshake, and asked, "Why didn't you tell me you and Col. Jacks are friends?"

"I didn't think it was important." I replied.

"Well, Col. Jacks got Personnel all straightened out. That's one battle I won't have to fight every time I need technicians!" Electronics was still a stepchild in the Air Force.

I was assigned as a radar technician in the radar jamming repair section, which was part of the Air Force's Countermeasures Radar equipment. It was easy to check out and to troubleshoot the radar equipment. We had hundreds to do and we were kept busy. After six months, I was made a supervisor over twenty-five male and female technicians, with a $1,000 a year salary increase.

My job was hard but I loved it and LaRue was so happy and thankful.

We began to look around for another house, one that we could buy instead of renting, away from the runways of Kelly Field. One huge transport plane had a problem on takeoff and veered over our house at about 100 feet, and crashed only a mile away. It did frighten my family!

We found a new, two-bedroom house in the Bellaire subdivision of South San Antonio, just two blocks from an elementary school and a half-mile from a large grocery store. It was a quiet neighborhood with nice people.

LaRue sold her old upright piano and I sold a shotgun and a Japanese rifle and bayonet to add to the few hundred dollars we had saved for the down payment. We still had a few thousand dollars guaranteed under the G.I. Bill that we hadn't used to buy the house in Bryan, so it worked out perfectly for us.

Later, I made it up to LaRue when I was able to buy the organ she had always wanted so she could continue to practice. She played either a piano or the organ at our Harlandale Southern Baptist Church each Sunday. I really enjoyed the church as well as the classical music she played at home.

I built an ornamental wire fence around our back yard and we had a garden and a nice place to have cookouts. Col. Jacks came out a couple of times a month and either had dinner or a cookout with us or took us out to dinner. We always had a great time together. We thoroughly enjoyed his stories about his employment as an engineer for the Stutz Auto Company.

He owned a “Stutz Bearcat” antique touring car, which he kept garaged at his family’s home in Indiana. He brought it to Texas for a few weeks and it was a thrill to ride in it!

My two daughters had only two blocks to walk to the elementary school and they both loved it there in the Bellaire Subdivision. We became active members in the Harlandale Baptist Church and made many friends.

LaRue was the Church pianist and taught Sunday school and I began teaching the men’s classes, which wasn’t too difficult. All the lessons for the teachers were nicely prepared and a format was provided for each teacher to follow.

With my experience in coal mining towns, and with mountain people, the Army Air Corps, WWII, and so forth, it was easy to recall some episode to use as examples to fortify the lesson being taught.

My men’s class donated fifty cents or more each Sunday to the “class kitty”, which was used to buy food for the poor families in our neighborhood at Thanksgiving and Christmas. Most of us gave quite a bit so that there would be frozen turkeys or hams to go in each box of food we gave. We were doing great until some of the recipients told us that their government assistance check was reduced the following month by the amount deemed they had received in food from us!

We were stunned and, much to our amazement, when we checked it out, we found it was true! It seemed that some jealous neighbors “turned in” the recipients to the government!

After that we made sure the “assistance” was less, which meant we could give to more people, and made certain that no one knew of the value of each bag of food they received.

At that time, I was smoking and drank a few beers now and then. I knew that I had to quit the “puffing” and slow down on the beers I drank at the Kelly Field picnics and such.

The four of us would go to Rockport to fish or camp our on the beach on Padre Island. We saw an ad for a lot for sale on the waterfront on Copano Bay, which is just two miles north of Rockport.

We talked to the realtor and ended up buying it with a small down payment and small monthly payments. The lot was covered with prickly pear and mesquite brush, but it was on the water!

I stopped smoking and started saving the money for the payments on the lot. In the meantime, LaRue was active in the Parent-Teachers Association. Our neighbors brought us their tiny kids whenever they had "business to take care of" for a few minutes, or hours, for free. With both our daughters in school, this was becoming a mounting problem and had to be stopped, due to the increase of the number of children being brought to our house!

The superintendent of the Harlandale School District, Dr. McCullum, called LaRue one day and told her that he had heard that she was a schoolteacher. He asked if she had a state teacher's license and which grades she could teach. LaRue told him that she did have her teacher's license and that she could teach any grade from 1st through 12th and that she was also certified to teach music and band.

Dr. McCullum then asked if she would bring her resume to the school headquarters the following Saturday at 9:00 A.M. LaRue went to the meeting and was hired as a music teacher. She was to start teaching the following week in two different elementary schools.

This was exactly what LaRue needed to stop being a stay-at-home mom for all the neighbors!

She taught at one school in the morning and the other in the afternoon. They put on choir shows for the PTA, and Christmas and Easter specials. The children all loved LaRue because she cared for each and every one of them as if they were her own.

One year, her class put on the "Sugar Plum Fairy" from the Nutcracker Suite. I was to take care of the recordings on tape while two of the mothers made the costumes for the dancers.

Each class put on a two-hour show, one at each of the school auditoriums, on two different Friday nights. Both auditoriums were filled, with people standing at the back. Dr. "Mac", the superintendent, saw one of the shows and asked LaRue to put on the show at the high school. The high school kids were really surprised and enjoyed it so much that they kept calling LaRue to congratulate her and thank her!

After several other music teachers were hired, LaRue asked to be moved to a classroom. She was being overworked and needed a change. Dr. McCullum gave her the second grade at the school nearest our home.

LaRue did a wonderful job as a classroom teacher. She was elected president of the Harlandale School District Teacher's Association, which had over 500 members.

She fought a successful battle to get the minority children breakfast at school along with medical care for those children with learning difficulties, through the Teacher's Association.

In order to get more funding from the federal and state government, LaRue had the Association sponsor a formal dinner for the teacher's and their spouses. She also invited members of the Texas legislature and the members of Congress who represented the San Antonio area.

There were special speakers and chamber music throughout the banquet. It was a huge success and won the hearts of representatives, securing more money for the district. One quiet little lady did make a difference!

One day, LaRue had to attend a staff meeting at 9:00 A.M. and asked me to substitute teach for her. To my delight, the first hour was reading and the second was art. No problems there! For the art class, I drew a picture on the chalkboard of a four-stack passenger steamer sailing on the ocean with the sun setting in the background and seagulls, represented by short wiggly lines, flying in formation over the ship in between the clouds.

The second graders, with pencil and paper, tried their best to duplicate what I had drawn on the board. Much to my surprise, they did quite well and enjoyed it very much. They all wanted me to come back!

The fallout from that art class was that every picture they drew—Thanksgiving, Christmas, and even Fire Prevention, had seagulls flying over pilgrims, Santa and his reindeer, and even over burning buildings!

"Joe," LaRue said, "I'm not sure I ever want you to substitute teach for me again! I've had to live with seagulls all year!" I was invited back a couple of times, but no more seagulls!

LaRue became a resource teacher, which meant that she planned and secured all sorts of training and visual aids and special methods for children with learning difficulties in the school district. Many returned years later to thank her again and again, which really cheered LaRue up and made her try to do even more.

In a previous chapter, I mentioned "more about George Palus later". We union officials of Local 5959 UMW had investigated an allegation that George had used union dues for personal purposes. He was found innocent of all charges.

Twenty years later and over 1400 miles from Caretta, WVa, where it happened, I was questioned about that by a federal agent. He found me at Kelly Field, San Antonio, Texas. Someone at Caretta had given him the wrong information.

I wrote a statement giving all the correct information which was typed and notarized and I signed it.

I found out later George was a Civil Service employee in the Pentagon in D.C. as a communication officer. He was applying for a crypto clearance which required a higher security clearance than he had. Much to my joy, he succeeded.

Out at Kelly Field, I was getting step raises and eventually a promotion in a way I never thought of before. While I was on my section of twenty-five technicians, I was kept quite busy getting them working smoothly as a team to repair radar-jamming units used to create "snow" on every radar screen near our section, part of the Enemy Countermeasures, our company's section. It was the Identification Friend or Foe radar.

We both had to meet production schedules each day in processing the black boxes to packing and shipping to Air Force Supply at Kelly AFB, where they were then shipped to whichever air bases needed them.

One day, John Gano, a planner in Electronic Engineering, came out and spoke as we always did near their office. The building had over 600 technicians in radar, 900 in radio, and about 50 in the Test Equipment Sector.

"Joe," John said, "How would you like to be a white-collar worker and help us in Engineering and Planning? I've seen your 201 file and we need you!"

I told John that I would have to speak to Tom Enderly, the Chief of Electronics, and would let him know. A few days later, I saw Tom. He smiled at me and said, "Yes, I know they want you in Engineering. I really shouldn't let you go but you've done more for this organization than I ever hoped for and this is a chance for you to move up the ladder in a new organization called Electronic Production Central. I'll approve a transfer and, Joe," He said, "you let me know if it isn't what you want. You can come back home!"

The famous saying at Kelly in the blue-collar section was: "I can't climb up the ladder of success because there are too many people stuck in the rungs!"

I was transferred to a planning position for long-range radar and gap fillers—all ground equipment for the Southwest area of the United States to be used against missile attacks from ships and submarines. John Gano and Art Immicke were my coworkers. It didn't take long for me to learn what needed to be done to keep those radar stations and sights functioning twenty-four hours a day.

I was given a GS-7 and got my step raises on time. My job required that I dress like and office worker since I attended a lot of conferences and meetings in the course of my duties each day. With my former military training, I knew I had to be professional. I had a secret clearance and the equipment such as aircraft radar, the ground radar technical orders (manuals) were classified.

One morning, the Branch Chief, Cecil Bain, came out to my desk and sat down near me. "I've been looking for a new Chief Security Officer to replace a man we lost." He said. "After looking at your military and civilian records, I think I've found the right man." He looked at me. "Would you be interested? Of course, it will be an additional duty but you'll meet some interesting people at the Directorate Security meeting every month."

I looked at him. "With a name like Diviak, are you sure you want me as Chief Security Officer?" The Cold War was in full swing back then.

Cecil Bain replied, "Joe, I know you're not Russian! I grew up with people of your parent's nationality and they are the finest and most honest people I have ever met. I want you for this job!"

I accepted the additional job, which proved to be an education in itself! There were a lot of perks, such as my own parking space, and I met a lot of Chiefs that later proved to be real friends. They recommended me for jobs that later aided in my promotions.

After some brief schooling, and after attending my first monthly meeting, I knew I could handle the job. After each monthly meeting, I had to hold our Branch security meeting. Once I got over the "shakes" of the first one, I did fine.

The first thing I did was talk to a man named Ben Jefferson, a black man over sixty years old who loaded radar and radio equipment packaged for shipment. He was loading the equipment onto a truck with a forklift. He looked under the truck and noticed a manual lying on the ground. He picked it up and gave it to the driver. Ben didn't know the manual was marked "confidential".

The truck driver, who wasn't cleared, turned it in to the Chief Security Officer, Ted Bush, my predecessor, a mean and unlikable man with a superior attitude of self effaced importance.

Bush immediately had Ben's clearance suspended and he was given a job outside the electronics building.

I took notes and wrote a formal letter through channels to have Ben's clearance reinstated. He had done what he thought was right and had intended no harm. Ben was due to retire soon, after twenty-five years of faithful work. Ben had no blemishes on his record before that incident and had received timely promotions from his supervisor.

About two weeks later, the Chief of Kelly AFB Security called me to his office. He commended me for taking the time to correct something that should never have happened to Ben. He told me the Commanding General had heard of the case and was very pleased with the results.

Security Officer Jack Price was happy that all had gone well with the reinstatement of Ben's clearance. Ben was given back his old job, much to his joy, although he wondered what had happened.

A few years later, Ben was retiring and as was the custom at airbases, you "ran a clearance", a one-page form with the key organizations, which you have to take to each department you did business with to sign off that you didn't owe them any money or have any of their equipment that you might have signed out.

Ben came to my desk and said, "Mr. Diviak, why didn't you tell me it was you who got my clearance back? I found out from Personnel that it was you." He came over to me and shook my hand then gave me a big "bear hug". "Now I can retire without a security violation on my record that may cause my family problems! Thank you!"

Ben moved to the Olympia Peninsula in Washington State to be near his son and his family. He sent me a picture of him with a salmon he had caught. It was very rewarding to see Ben so happy, with his family around him, enjoying his retirement.

To keep the early warning radar and gap filling radar operating in the Southwest, the parts needed and the technicians necessary had to be preplanned weeks ahead for the preventative maintenance.

I spent days scanning the technical orders (manuals) on each system and singling out the parts that would most likely need to be repaired by part number and stock number, along with parts maintenance records. It was still a job to secure the parts in supply for the teams servicing the area radar set up on mountains in Arizona and New Mexico. But with a lot of luck and careful study, we managed to do the job well.

We knew that high-wear items, such as motors, dynamotors, megatron tubes, and other vacuum tubes would be the parts most likely to need repair.

The British had invented radar before WWII and they came up with better radar that could cover over a thousand miles with one set, which made all the earlier models along the coast and on the mountaintops obsolete. It was a fantastic improvement and yet so simple to do, once the secret method was given to the electronics manufacturers.

My new position was to be assigned to the radio section of the airbase electronics. This branch was expanding because all the new aircraft had new systems that had to be planned for and we still had to maintain the old systems.

Gentile Air Force Station in Dayton, Ohio, was the prime depot in charge of getting all airborne radio and radar equipment repaired and to purchase new units with all the cabling and test equipment to do the job.

We were the support organization and our huge two-story building was constructed with metal and other screens to prevent any beams or electronic waves from interfering with the radios and televisions in the community off base.

On a Monday morning, my supervisor gave me a briefcase containing a lot of information that I would need and travel orders to go to Gentile the following weekend to secure the workload for the radio shop for the coming fiscal year.

My fishing buddy, Albert Carter, was given the same for the radar. He had made the trip before and clued me in on what to expect at Gentile.

I was stunned; where would I start? I began by getting the flying hours each aircraft had flown the previous year and what systems each had aboard. That way, I could get the number of black boxes in each system and we had a failure rate for each system.

Then I went to the shops to see how many mockups (test stations) were on each production line, and most importantly, the maximum production rates. That, coupled with how many man-hours each employee would be available to work for that year, annual leave, sick leave, jury duty, military reserve duty, and so forth, had to be factored into the totals.

I really began to appreciate the fine math courses I'd had in Big Creek High School! The teachers there were terrific and I wrote them letters of appreciation many times later.

When I had gathered all my information and Albert had gathered his, we found that we needed to secure 1.5 million man-hours worth of work for our huge repair facility for the upcoming fiscal year. It broke down to nine hundred thousand man-hours for the radio shops and six hundred thousand hours for radar.

A "slippage factor" of ten percent more man-hours than we actually needed was built into our forecast. This would cover late receipt of black boxes from overseas airbases, etc. The Aircraft Division always had to borrow electronic technicians because they always miscalculated and ran short of workers to complete their repairs. Our technicians could easily work on aircraft but the aircraft technicians knew nothing about electronics and didn't want to know!

Finally, Albert and I were ready for our first trip to Gentile. We would be competing with manufacturers and two other Air Depots for the workload for the next fiscal year, which was only six months away.

We flew by commercial airline from San Antonio to Dayton via Chicago in an old propeller-driven aircraft. It was an all day trip! We had a reservation at a hotel and the hotel limo picked us up at the airport.

After breakfast the next morning, a staff car picked us up and took us to Gentile. We were shown into a conference room and met a lot of nice, friendly people.

In five days, everyone had secured enough man-hours to satisfy their home facilities and I was elated that all had gone so well. From the outset, two people we worked most closely with there had been very friendly, even inviting us to their homes for dinner.

When they traveled to visit us every two months, I always picked them up at their hotel in downtown San

Antonio early in the morning and took them to a café near Kelly Field for breakfast. They were exposed to ham and eggs and red-eye gravy, and hot biscuits, or sausage gravy on biscuits and Western omelets—all good Texas-style cooking! They loved it and survived and thrived on the food.

I would take them to my home for dinner or a cookout one evening during the week.

We also traveled to Dayton often to adjust the current workload and work out requirements for new systems on new aircraft coming into the Air Force.

Our competitors, the manufacturers, bid on repair jobs but when they were given a large contract, they would "cannibalize"—take parts from one black box to repair four or five others and throw the "carcass" away. The government shops repaired every black box so we wouldn't have to purchase new ones.

In between all this, we had to attend parts provisioning meetings regarding new radio and radar systems that were being installed on the new aircraft coming into the Air Force. I always took a shop foreman with me because they knew what their technicians could repair.

The manufacturers were Collins Radio, RCA, G.E., Hoffman Radio, Bendex, Halacrafter, Magnavox, among others. All were located in different states.

We usually included a couple of people from Supply and some management people from Gentile in Dayton. The trips were very informative and gave us an opportunity to plan ahead for test mock-ups and cabling. This gave Supply a chance to get all the parts we would need in place. The trips usually lasted from three to ten days. The first thing we did before we went into a conference was to tour the plant to see the new equipment being manufactured.

Before we processed our travel orders through Base Transportation at Kelly Field, we had to read and sign a certificate stating that we would not accept any gifts or gratuities from the manufacturer, including food and drinks, etc.

That was fine with me, but it did infuriate me to read the newspaper stories about our congressmen, who are government employees too, accepting free vacations to all parts of the world, and golfing trips to the Caribbean and other places, from the parasites representing wealthy clients hanging around the Capitol, all of whom needed special favors to circumvent the laws that were in place. We felt that this was a double standard, of course, put in place by the very people we voted for to uphold the laws of our country!

We "peons" still managed to get around this. When we ate in the executive dining room, we would ask the cashier to give us a bill marked "paid" and would hand her a dollar bill—we had "paid" the bill and we could swear to it!

Our travel orders didn't show our security clearances, but before our departure, messages regarding our clearances were sent to the factory. They were really great in maintaining tight security and only gave us what we "needed to know" to complete our preplanning at the conferences. We were representing our Air Force facility as civil service employees and any violations could cost us our jobs. Without it, you couldn't work in electronics.

This was a constant concern of mine since I was a Chief Security Officer in addition to all my other duties. One mistake and my whole career could go down the drain. The pressure was always there.

In the midst of this hectic schedule, Wilbur Thompson, a friend who was working temporarily in Directorate Headquarters, intercepted a strange message. He brought it to me to read because I knew all the culprits.

It was written, in my opinion, by an unlikable supervisor, Bill Dorman, at Gentile Air Station in Dayton, Ohio. He apparently wanted to build an organization at

Warner-Robbins Air Depot in Georgia and eventually eliminate our facility at Kelly Field.

It was a lengthy letter about how Dorman would move to Warner-Robbins as Chief of the new organization. It had some high-ranking approval and was probably sent by a friend in Gentile Station to the wrong address at Kelly on purpose.

Wilbur showed the letter to Henry Champion, a radio shop supervisor and the shop steward in the American Federal Employees Union. Henry decided to write a letter about this to our congressman, Henry B. Gonzales.

He brought me the rough draft to check to make sure there were no security violations. I read it over and had him take out a few paragraphs dealing with personnel problems with the Aircraft Branch. I told him that this was not the time to air our dirty laundry; we should focus on this one subject and keep the letter brief.

About the same time Henry Champion was writing his letter, our on-base newspaper, The Kelly Field Flying Times, came out with an article on whether a federal employee could write his congressman without fear of losing his job. The article quoted the Commanding General as saying it was a privilege, so long as it didn't contain any classified information. Of course, we all saved the newspaper with the article!

Much to my surprise, a copy of the article was sent to our Senator Yardbourgh, who none of us knew or had ever met. When the Senator received the article, he immediately sent it to the Secretary of the Air Force.

A couple of weeks later, Henry Champion called me to tell me that the three of us would be "on the carpet" soon and to be prepared. I was on my home phone and after he discussed a few things, I asked him not to say anything more on the phone to either Thompson or me. I felt we were being "bugged".

A Major Nunez from Security, Wright-Patterson Headquarters in Dayton, Ohio, asked my supervisor if he could talk with me privately at base headquarters, that it was a confidential matter. I met him in a room that had only a table and two chairs. I knew the room was bugged because, being a security officer, I had heard about it.

The Major was a crafty individual and he tried his best to shake my story that I did not personally write the letter. He used phrases that Henry Champion and I had used in our one and only phone conversation on my home phone.

I told the Major that I had nothing to hide and that the letter I had seen from Gentile didn't have "confidential" or "restricted information" stamped on it, which was required if it was not intended for strict privacy. He left a little upset, which made me feel better.

About two weeks later we were called to the Chief of Personnel's office and given a note–not a letter—that we were on "30-day leave of absence without pay" because we had violated "preplanning confidential information to persons not entitled to the need to know only" regulation.

What stunned me the most was that they did not deny the underhanded work they were planning, to set up an "empire" at another depot at taxpayers expense, when there was already one doing a tremendous job at Kelly Field!

The suspension didn't worry me too much because I knew it "wouldn't fly"!

We were confident and immediately arranged a meeting with our congressman, "Henry B", at his office in San Antonio. He couldn't believe our notices and said, "Go home and I'll take care of this. It's ridiculous!"

We all knew that Henry B was a real fighter for rights and a man of his word! I voted for him when he ran for City Council years before, when he was just starting out in politics.

On the tenth day into our unpaid leave, and after our names were splashed all over the two newspapers in San Antonio, we were called back to personnel and reinstated without any pay loss and our personnel records were cleared of any wrongdoing.

Congressman Gonzales had the whole bunch of people at Kelly, Gentile, and Robbins Air Base scrambling to cover their tracks and worried about their jobs!

The Commander's message to the San Antonio newspapers was "these employees thought they had acted in the best interests of their airbase and that a lesson to be learned was not to talk out of school about matters in preplanning stage."

That really made us angry! We sent a copy of that article, along with the one from The Flying Times that quoted him saying it was "a privilege to write our Congressman", to the Commanding General of the Air Force in Washington, DC and a letter stating that we felt this was a very inappropriate statement. We also mentioned that we had a lot of support in the civilian population and they wanted to know "if these three men hadn't intervened in the preplanning stages, would it not be too late after it was all finalized?"

We were reinstated to our jobs at full pay. Three weeks later, the three of us "culprits" went to Personnel to look at our 201 files, to be sure our records were clean.

We went at lunchtime when we knew there would only be one person on duty. Low and behold, there was a letter in a manila envelope in each of our files stating that we were "untrustworthy" or words to that effect!

We promptly reported this to Henry B and it was promptly taken care of! I felt sorry for the clerk at Personnel. She was supposed to pull the manila envelopes before she let us see our files. She wasn't briefed beforehand. As Security Officer, I was aware of the envelopes because I had to check each prospective employee's file in our branch before I applied for their security clearance.

Our "good buddy", Wilbur Thompson, was at a bank near Kelly Field a few years later. While he sat in an office, he overheard a remarkable conversation between a bank employee and a customer in the next office.

"How are you, General Omar? Long time, no see since you retired!

"I retired before you did, Colonel Buckley, and as vice president of this bank, I've been very busy."

"I heard that you didn't receive the other star when you retired," Colonel Buckley said. "What the hell happened?"

"Oh, it was one of those problems where you have three civilian SOBs talking out of school!" Omar replied. "They got a bunch of politicians involved and I was the scapegoat!"

"It happens," said Col. Buckley. "When you receive a little adverse publicity under your command, the Chiefs up the line all scramble to cover their butts and leave you to take the blame!"

When we heard this, the three of us were quite happy and felt we had done our best. "The chickens had come home to roost". Later, the newspapers said that we had "saved hundreds of jobs at Kelly Field at the risk of losing our own jobs." Of course, it was printed in the back pages of the newspapers, somewhere near the obituary column!

After Gen. Omar had retired, the three of us were cautioned to be careful because there were a few civilians and military people around who didn't take too kindly to what we had done to "their" General. They still felt an alliance to him for the jobs they held due to his administration. I already knew that and had warned my two friends long before we received that advice.

We were a bit apprehensive about our chances of promotions and step raises. We figured they would all be a little slow in coming. The worst part was that some of our friends and coworkers kept their distance because they had been advised to. It seems that even in those days "you were judged by the people you associated with" on and off the base.

A few months later, the three of us were at the Kelly Field Annual Picnic. We were at a table set up under the live oaks, celebrating with wives. There was a lot of food and beer and a couple of hundred people there.

One loudmouthed individual who had a lot to say and too many beers came over to our table. "Boy," he said, "WE really saved a lot of jobs this year! I was right behind you all the way!"

I replied, "You must have been a long way behind us because I didn't see you!" He left grumbling to himself.

It reminded me of some of the coalminers I knew when I worked in the mines. They never attended a Union Meeting because they were afraid the boss would see them going into the meeting building and others who got deferred from attending the meetings because they held company management jobs.

After they retired, they enjoyed the health benefits, black lung checks, hospitalization insurance, Union retirement system, and burial insurance for their families, all because of the United Mine Worker's Union. We fought for it while they did nothing to help us, yet they always said, "WE had a tough battle, but we won!"

A few years later, with a new Commander and new civilian chiefs, the atmosphere changed. I was still Chief Security Officer of my branch and alternate to the Directorate Chief, Mrs. Lucas, who was a fine lady and did a magnificent job for many years before she retired.

Mrs. Lucas defended me against all criticism about the "Kelly Field incident", as we later called it.

Our trips to Gentile became an every-other-month deal. Albert and I were GS-7's and we were dealing with GS-9's and GS-11's, yet our job descriptions were the same. Why? We were told that living in the south was cheaper than in the north! That was a farce. The GS-9's were in the Division offices and we were doing their work for them because they knew nothing about electronics. There was a lot of politics involved.

On every trip, at every meeting, we carried as much authority as the people we were visiting. Why? Because we represented our Commander, who was a two-star general. The decisions we made at those meetings were valid and we were treated with respect.

Eventually we received our step raises and then our GS-9's, which called for a picnic at Kelly Field Park. Several of us who were being promoted purchased 16-gallon kegs of beer, which was iced down and delivered by the brewery, along with pumps and paper cups.

I went to the tamale plant and picked up two five-gallon cans, twelve dozen to a can, of steamed tamales, chips, etc. We had a glorious party on a Friday afternoon after work.

Many of our Chiefs were there and enjoyed all the kidding and congratulated each of us. I felt better all over after that.

I knew then, to use an old coalminer's phrase, that I wasn't "blackballed" because of the "Kelly Field incident"! Those picnics and promotion celebrations did more to solidify the Kelly Field workplace than anything else. The workforce consisted of people from all parts of the country and of all nationalities. Regardless or color or heritage, we respected each other and enjoyed a great working fellowship in our directorate of 9,000 civilians.

The civilians at Kelly Field had a Welfare Fund where twenty-five cents was deducted from our paychecks each month. The fund was used to send nice floral sprays or wreaths whenever an employee lost a family member, such as a parent, a spouse, or a child. We all felt this was a nice thing to do.

When my mother died in 1956, we found a nice, elderly lady who came to our house to stay with our daughters while LaRue and I drove to War, West Virginia. It was a long, hard trip but we arrived the evening before the funeral.

We checked into a hotel then went to my parent's home. Much to our surprise, my mother was in a beautiful casket in the living room, surrounded by beautiful sprays and wreaths of flowers. It shook me up a bit until I saw all the elderly retired people there.

It was more convenient for them to have her there at the house because the funeral home was in Welch, fifteen or twenty miles away, a good forty-five minute drive through the mountains.

There was a large floral wreath from Wright-Patterson Field in Dayton, Ohio, where my brother Steven worked. There was another from Jacksonville Naval Base, where my sister, Josephine, worked. But there was none from Kelly Field! LaRue and I had a nice floral arrangement sent from Bluefield, Virginia, which was there.

I was very angry but had no choice but to wait until I returned to San Antonio to find out why there were no flowers from Kelly Field.

Once back at Kelly, I questioned our secretaries about it and they said, "We only send flowers if the parents reside locally." Which a good ninety percent or more did.

Then I asked why the twenty-five cents was deducted from my paycheck every month if I wouldn't share in the benefits! They couldn't answer that one.

I wrote a letter through channels, over my supervisor's objections, which shook up the whole program. I received a belated letter of apology and the offer of a potted plant. I declined the plant and told them to stop deducting money for flowers from my paycheck. Of course, they didn't.

When my father died three years later, in 1960, the same thing happened and once again, there was a lot of "buck passing". I kept the "kettle boiling" though.

A well-dressed gentleman from Personnel came to see me privately. He asked me to let up on the letters and told me that a lot of people had been reprimanded and letters placed in their 201 files.

I asked to see the letters and he reluctantly showed them to me. "Please, Mr. Diviak," He pleaded, "Let it die!" What an untimely phrase! But he breathed a sigh of relief when I agreed to let the matter rest.

As we left War after my father's funeral, we drove through Caretta. The old tipple was gone and so were all the tracks. Most of the houses were gone, too. I took a picture of the Caretta Shaft Mine through the windshield of my car. It had been stripped of all the equipment. Just a few houses remained. It was a sad feeling to see my old home as a ghost town.

Caretta was a shaft-mine ghost town in the 1960s.
This was before strip-mining of the mountain tops down as much as 125 feet to get to the three-foot seam of coal, which was at the same elevation above sea level in all mountains there.

A short time later, my supervisor sent me to a meeting at Headquarters concerning the use of the Funds Surplus. I did my homework well. I knew my supervisor was an avid golfer since we fished together on the coast. I had heard rumors that "the fund" would be used to underwrite some of the expenses of the nine-hole golf course located in the center of Kelly Air Force Base, if civilians would be allowed to use it.

At present, only officers were allowed to use the course, which had been built by enlisted men during WWII and was later taken over by the Officer's Club. It was a nice flat course with a large clubhouse, a pro shop, and a bar, often referred to as "The 19th Hole" by the golfers.

I had noticed as I drove past it on my way to and from work that very few people used the course.

I waited until last at the conference to offer my suggestions. "Since most of you favor "The Fund", secure permission for civilians employed on base to use the golf course." I said. "If that isn't approved, I have an alternate suggestion. Dig up the golf course, to a depth of four feet, down to the clay strata. The dirt can be used to fill in the huge gully behind Kelly Field." I looked around the room.

"Then fill in the basin with water from Leon Creek, stock it with perch, crappie, and bass. The clubhouse would be the concession stand where canoes and life jackets and boats could be rented. That way a lot more people could enjoy it instead of just a privileged few."

Everyone had a good laugh. I returned to work knowing full well that as long as the officer's had control of the golf course, it would never change.

A few days later, my supervisor called me into his office and, to my surprise, showed me the Kelly Field newspaper: "A suggestion was made to turn the golf course into a fishing pond by a representative from Electronic Engineering

Planning. Joe Diviak's suggestion made about as much sense as the rest that were offered at the meeting."

My supervisor was grinning at me. "Why in the hell did you do that?" He asked.

"I did it in fun," I said, "Because I could see the officers had already told the civilian representatives it would retain 'status quo'! No changes!"

I took a lot of kidding for months after that, and some compliments too. I thought the number of conferences I would attend would diminish after that, but instead, they doubled! Many times, they asked for me by name. That made me feel a lot better and I really enjoyed myself. Much to my surprise, I was treated with respect and received quite a few letters of commendation, which helped me get my promotions each time I became eligible for them.

The Branch Chief came to my desk one day. "Who do you know at 'the palace' (headquarters)?" He asked.

I looked at him, clearly puzzled, and asked, "Why? What happened?"

"I received a call from a lady who is in charge of the Suggestion Awards staff at HQ," He said, "And she asked that you attend the next meeting, which is a week from today."

"That's news to me," I said. "But I'll be glad to go if you wish."

"You'll be there, Joe!" The Chief replied.

Before the conference, I went to our Division's Chief of Suggestions and asked for any interesting or outstanding suggestions his committee had pending. I made a lot of notes, concentrating especially on the Colonel's and Deputy Director's directives, etc. It took me about an hour.

I was at the conference twenty minutes before the meeting was to begin that Friday and met Lois Daley, the chairperson who had asked for me to attend. We became instant friends and, in my opinion, she was the most informed and best-educated person on base!

We spent about two hours reviewing the high-value-saving suggestions one by one and cleared the backlog. Then I had three that were pending from our Division. They were all approved and action taken to submit them to Finance for consideration.

For four years, I attended all the meetings. Lois Daley was one of my best friends on base until she was promoted to a higher level.

She would often call to discuss problems that she encountered. She knew that I had been on Kelly Field for years and knew a lot of people.

The Suggestion Awards Plan was a huge success at Kelly. It saved the government millions of dollars each year.

I kept wondering who had given her my name. One day, we were attending a management meeting together and I asked her. She told me she had gotten my name from a mutual friend, a high-level civilian that I had helped out of a problem with security. She said this person had told her, "If you want someone from Airborne Electronics Engineering, try Joe Diviak. He's knowledgeable on many subjects and very trustworthy."

I knew where all the rocks were! It seems that all my life, I did the best I could for my family and myself, but I never forgot to help others along the way. I did this freely and without expecting anything in return. I was rewarded sometimes even more than "seven fold" later in my life. It's the most amazing thing!

Our church had an old school bus with a worn out engine. At one of our monthly business meetings, it was suggested that we sell the bus to the highest bidder; then the church could buy a larger bus. We had over seven hundred members and were still growing and needed one.

I submitted a bid for fifty dollars and to my surprise I got the bus! I towed it over to a shop and had an overhauled used engine installed in it. Then I parked it beside our house in Bellaire so I could junk the seats.

I installed a new wall-to-wall vinyl tile floor, intending to make a camper out of it but it was never used for that! About that time, our Church wanted to expand the Sunday school with a new, two-story building with ten classrooms, complete with a full-size basketball court on the flat roof, enclosed by chain link fence and equipped with lights.

An elevator would be used to get from the first floor to the roof, along with an inside stairway.

In less than a year, I helped sell over $750,000 worth of bonds at six percent interest that would mature in five years.

In order to make room for the new building, the old log cabin-type Boy Scout building would have to be torn down. I bid on the cabin because it was made out of 7-foot long used railroad ties, and I needed 62 for fence posts on our waterfront lot on Copano Bay.

I won the bid and we tore it down. I was able to sell the rest of the ties and hauled the "junk" that was left to the landfill. Those ties made a fine fence, with five strands of barbed wire and a locked gate, around three sides of the property. The beach side was left open.

LaRue was offered a higher paying teaching position on the north side of San Antonio by a principal she had worked with before. We bought a nice large house near the school and sold our old home easily.

The bus came in handy for moving most of our household items and clothes, etc., but a commercial moving company moved the furniture and appliances.

The new house was a dream! It had over 1,800 square feet of space and a 2-car, attached garage, and a large lot in a quiet neighborhood. It was a great location, near two shopping centers and easy access to all the major highways looping around the city.

I measured the back yard for a privacy cedar fence with two large access gates. It would take 320 feet of lumber six feet high to enclose the yard. We took the bus and went to a cedar sawmill in Bastrop where we purchased the fencing boards and posts we needed.

I built the fence myself, working evenings and on weekends. It really looked great with the new trees we had planted back there and I had a place to park the bus out of sight of the street.

In between all this working, we still managed to take trips to LaRue's parent's home in Corsicana as well as a few trips to the coast where there were a lot of motels with little kitchens. I trailed our little boat and motor and we fished the back bays. If anything happened, all you had to do was stand up and walk back to shore. It was only four-feet deep for miles. But there were a lot of fish there! LaRue and our daughters, Ellen and Barbara, became great fishermen!

We switched our membership to Castle Hills Baptist Church, only two miles from our home. It was a very active and growing church.

Our Pastor, Jack Taylor, had a couple of small children around ages five and seven. When the riding stables near Lackland Air Force Base decided to sell a small Shetland pony, Reverend Jack bought it but had no way of getting it to the farm of one of the members just north of the city.

I drove our bus to the stables and picked up the pony. I loaded him through the back door of the bus and tied the bridle reins to the bar behind the driver's seat, and we headed for Castle Hills. It was a Saturday afternoon and traffic on the Loop 13 was a bit heavy. Apparently the other drivers could see the pony's head over the steering wheel but couldn't see me. Some of them even turned around and began to follow the bus! They had never seen a pony driving a school bus before!

Finally I backed the bus into a grassy bank and unloaded the friendly pony. He wasn't a bit nervous!

On Monday morning, my supervisor came over to me. "Joe, I saw your school bus on Saturday, but a pony was driving! Did you have some sort of costume on?"

"No," I said, "But the pony learned to drive with just one lesson!"

It was easy to spot my bus. It was yellow around the sides, front, and rear, but had a snow-white top, down to the tops of the windows. I painted it with a spray gun, after I had knocked out all the dents.

We also used the bus to take children to various church encampments, choir contests, and cookouts on the Guadeloupe River in the hill country. The boys were always willing to put wooden pews in the bus for the trips and remove them and put them back in storage when we got back. The children loved it!

Early one Saturday morning, an elderly gentleman came to our door. He wanted to buy our bus. He also had a nice 35-foot aluminum mobile home for sale at a nearby mobile home dealer's lot. Since we had our lot on the Texas coast and the mobile home would be perfect for that, we swapped the bus for the mobile home. The mobile home dealer towed the trailer to Rockport and set it up on our waterfront lot.

Our church had purchased a new larger bus that held eighty passengers and was air conditioned, so they didn't need my bus.

The mobile home was well furnished except that it had a gas stove and gas hot water heater, which I replaced with electric units. It was comfortable for the four of us and looked great on the shore of Copano Bay. We were just fifty feet from the water and it was great for wading or boat fishing.

I installed an electric meter loop on a twenty-foot telephone pole beside the mobile home. I put a large outdoor mercury vapor lamp on the same pole to light up the place every night.

After putting piers under the trailer, which raised it three feet off the ground, and anchoring it firmly with large screw anchors (like the power companies use to hold the guy wires from their poles, fastened to the trailer's mainframe), the trailer was there to stay. It went through two hurricanes without damage!

The hardest job was putting in the septic tank and the drain laterals to accommodate the sewage outlet from the mobile home. I had to dig the entire thing myself but eventually it was finished and passed the County inspection.

We had to haul all our water in five or six five-gallon cans to fill the reservoir on the side of the trailer. The city had a park on the beach complete with showers and water faucets, so there was plenty of water available. There were plans to extend the water main about half a mile, to our subdivision, in a couple of years.

The four of us spent many weekends, holidays, and vacations in the mobile home. Because the property was on the bay, it was great for wade fishing. We went beachcombing and did a lot of sightseeing along the Padre Island shoreline. Corpus Christi was a nice town to visit with great restaurants.

We enjoyed being in the surf, then taking a freshwater shower before we relaxed for the evening. Being a teacher, LaRue enjoyed the break from the every day routine, as did our two daughters. I was able to set aside all my problems while we were there and we all relaxed and enjoyed it. After all, I couldn't worry and fish at the same time! I felt rejuvenated after each trip to the coast.

After three years, we started looking for a place around Aransas Bay, in another part of Rockport, where the mosquitoes were not such a problem. The area was sprayed once a week during the summer months by a mobile unit that drove along the shore road. There was always a prevailing wind there as well.

My brother-in-law Don, and his wonderful wife, Betty Ann Cummins. They are my best friends and pals since 1942. They live in Austin, Texas. They are family.

My brother-in-law, Dan Cummins, and his brother-in-law, Tom Bradfield, and Tom's father, Landon, owned a real estate firm in Austin. They had a nice cottage on the Aransas Bay Peninsula, which had only one row of homes along the road that went to the end of the peninsula.

They also owned several vacant lots between the houses and had a deal to buy two small houses next door to their cottage and a vacant lot where a cottage had been demolished by a hurricane.

The Cummins Bradfield Real Estate Company purchased the properties and offered me a deal to purchase the cottage next to theirs. It was a dream come true! I sold the mobile home and lot for more than twice what I had paid for them!

The cottage was on a waterfront lot, about a hundred and twenty miles from San Antonio. It had two bedrooms, two baths, a large living room, a dining area, and a long glassed-in room on the front. There was a nice kitchen and serving bar. As many as sixteen people had stayed there at one time. There was a boat basin with covered boat slips and a 1,000-foot wooden pier with lights for night fishing.

The cottage itself needed a lot of cleaning and painting inside and out. The former owners, two families, were very generous about giving their friends keys to the place. Of course, their friends loaned their keys to their friends, and so on. By counting all the different sizes of wading boots and clothing, we figured there were at least forty-eight people using that cottage!

It took five pickup truck loads to haul away all the old clothes, bedding, shoes, broken fishing tackle, pots, pans, and spoiled canned goods in the cabinets and food in the refrigerator. The worst was all the medications left in the two bathrooms. There was everything from treatment for dandruff to ingrown toenails, and all the body parts in between! We sure were glad there was a bulldozer at the landfill where we took the trash!

After days of cleaning, we finally had the house in a livable condition. We changed the locks on the front and back doors and posted a notice on the front door: THIS HOME IS UNDER NEW MANAGEMENT. Even then, a few times while we were there, a car would pull up and someone would try to open the door with their old keys.

There was a ten-foot wide and twenty-foot long cross pier at the end of the long pier, which gave us more fishing room in the deeper water. The lights were old floodlights and being an electrician, I shopped for more powerful lights that would draw more fish to the pier.

I bought three 500-watt lights with built-in reflectors and mounted them on eight-foot high pipe poles. They worked exceedingly well. We caught fish every time we fished out there after dark. We caught flounder, speckled trout, white fish, drum, croakers, golden croakers, red fish, and gaff top. We set traps for blue crab and had all we could eat! It was a fisherman's paradise! We always shared our bounty with our relatives and neighbors in Rockport.

We would clean the fish then fill half-gallon milk cartons with fillets and water, seal the lid flaps and freeze them in our deep freeze. It was a perfect way to have fresh fish during the winter.

In the twelve years we owned the cottage, it weathered five hurricanes with only minor damage.

I built drop-down awnings out of plywood on all the windows on the front and south side of the house. They could be propped up to keep out the rain and sun when we were there. We would drop them down over the windows and fasten them when we left. That way, we never had to rush down there to board up the house whenever a hurricane popped up in the Gulf of Mexico.

A constant southerly breeze of the water that kept the cottage cool. We had no air conditioning nor a need for it. It was a wonderful place for a nice nap or read a book or to write a letter on a summer afternoon.

I installed a freestanding fireplace in the middle of the large open area. We would burn driftwood, which always had all sorts of chemicals from the salt water and would spurt out different colors as it burned. It also supplemented the electric heaters we used in the winter.

It's strange how our popularity increased with people who were distant in their friendship before. Suddenly, they were all warmth and charm, hoping for an invitation to the beach house!

Some of them became angry because we wouldn't let them use the beach house for a weekend. A couple of supervisors at work wanted to rent the beach house but I refused. This caused a little unwarranted bitterness toward me but I knew that if you gave someone the key, next thing they would do is have copies made and share the fellowship with their friends!

Most of those people had children who could easily wreck the place in a weekend! We had enough problems keeping the plumbing working right and the place secured against vandals. We had a neighbor two houses from ours who would "keep an eye on the place." They wouldn't take any money for it so we would often buy them gifts to thank them.

My brother-in-law and his brother-in-law and their families had a beautiful home next door to us and one or the other would go there often. The pier had a locked gate. It wouldn't have been fair to them to allow other people go there with their beer and their mean kids. It was a very quiet, clean area and most of the ones who wanted to "rent" the cottage weren't interested in fishing. They were more into partying with loud radios.

We had our own parties but we never intruded on our neighbors. We enjoyed many, many wonderful weekends and vacations there.

We became acquainted with the neighbors and, after a hurricane or maintenance problem, it was my pleasure to help them with repairs.

I kept the grass mowed because of the snakes and scorpions seemed to love damp places. I also installed a large mercury vapor light on a thirty-foot pole across the street from our cottage. The light had an automatic daylight and dark control and lit up the area in front of our neighbor's house as well as ours. The neighbor appreciated it. There were no more break-ins!

Back at the air base, being a supervisor required all kinds of training sessions every time a new set of regulations had to be "implemented", a much over-used word in government service!

One of the courses was on how to boost the morale of the workers putting in long tedious hours of repetitious work. It was felt that awarding advanced step raises or Superior Performance Awards would accomplish this goal.

One of my fellow supervisors asked Mrs. Lois High, our Chief of Personnel who was teaching the course, "What do you have to accomplish to receive a Sustained Superior Award?" This award was a nice bit of cash.

Mrs. High replied, "You have to be perfect in every way!"

Someone in the back of the room said, "I read somewhere in a book that there was one guy like that. They beat the hell out of him and hung him on a cross! I'm sure I don't want to be perfect!" Everyone burst into laughter, including Lois. She dismissed class for the day right after that.

The next day, she apologized for her flippant answer and told us that she appreciated a class with a great sense of humor! She was a wonderful lady and was in charge of thousands of employee records on the base.

The perks for this type of award was a few hundred dollars and a certificate was placed in the employee's 201 file. It would help them to receive their step raises in pay several months ahead of time.

The downside of this was that a technician had to wait until the chiefs awarded all their secretaries and favorite employees first before it eventually trickled down sparsely to the people who really deserved it. The technicians who did all the work on the black boxes in electronics, sheet metal, accessory repairs, engine repairs, etc.

Being perfect to an extreme was not a desirable quality on a production repair line. The flow of materials and repairs was coordinated by studied labor standards and any unnecessary time taken to satisfy a person's "perfection phobia" would ultimately cause delays and loss of production.

We did excellent work and every black box operated according to the technical order book written by the manufacturer to set the pulse, wave forms, etc. When it went to packing, it was working to the manufacturer's specifications.

I always had the strange feeling that a perfectionist had a deep-seated inferiority complex and was afraid that someone would find out. They just had to be perfect.

The time came for me to attend a five-week course in Logistic Management for supervisors at the Air Force Institute of Technology at Wright-Patterson in Dayton, Ohio.

I stayed at the Visiting Officer's Quarters while I attended the classes. Each class had twenty supervisors from all over the country. We had twenty-three instructors, all with doctorate degrees in all the fields dealing with people and some had as many as three PhD's. I looked them over and knew that we were in for a hard five weeks of work!

Of course, we had the usual "self-important know-it-alls" in each class and it soon became evident long before the classes even began that most of them didn't feel they should be there.

But I liked the speech our class's professor, Dr. Bill Godown (PhD three times!) made. "You were all sent her for a purpose," he said. "To learn to handle not only supplies for aircraft, but also to handle people fairly and equally for maximum production. However, anyone in class who feels he does not need this training may leave now! No problem." He looked around the room. "But, of course, we will send a message to your commander telling him that you know it all and that he made a poor choice and wasted government money sending you here. Any questions?"

Dr. Godown looked around the room again. "None? We will start now!" Dr. Godown, a psychiatrist, then proceeded to hand each man in the front row seats a "buckeye nut", a horse chestnut, and he kept one for himself. Those of us who had received a nut were the appointed monitors. If he got off the subject in discussion, one of us was to hold up the horse chestnut. When we did, he would hold his up, too.

We used those horse chestnuts a lot! And we were all keenly aware that the horse chestnut was a silent and sanitary expression of what a horse produces in abundant quantity and frequently!

Those were the busiest five weeks of my life and time passed very quickly. We had classes all day and we had to read three books on human behavior and give a twenty-minute report on each one to the class.

BUT, I learned a lot that was really beneficial in handling people and being a better man emotionally.

Another course on the management of material in a timely manner to support overseas squadrons was another great course. Part of what we learned was how expensive storage space was per cubic or square foot to store over-buys, etc.

All the professors gave talks in our class about how to spot problems in employees as well as supervisors. Several classes attended these talks.

One of the many subjects covered, the one I liked the best, was Human Relations and How to Get Along With Each Other. I still use what I learned in that class to understand others who have problems.

After graduation, we had a farewell party at a nice nightclub, in a private dining room, along with the other classes that had graduated. During the party, Dr. Godown and I had a few drinks together and a long talk. He was quite an interesting person.

He told me that one of the theses he wrote for his doctorate was called, “The Law of Probabilities”. For instance, he got a bunch of fellow college students to throw dice to see how often a seven or eleven would come up in 3,000 throws. Another experiment was the chances of filling an inside straight in poker, and many more.

After he received his PhD and graduated, he took a trip to Las Vegas and won enough money to buy a new Buick, spend a week in Frisco, pay all his expenses back to Dayton, and still had $10,000 left!

I asked him if he was planning to go back to Las Vegas. “No!” He replied. “In all probability, I would lose my butt!”

After graduation, we all returned to our homes and bases, a bit wiser, knowing that we could do better with the people we worked with in our jobs and spoke to by phone across the country.

I returned to my many jobs and worked hard to make sure that everyone I worked with was happy and producing a good product in the shops. I had the responsibilities of being a planner, work loader, security officer, and overseeing the overhead budget, plus the support groups for the shops!

One morning as I was making my usual visit to one of the shops, I saw one of the foremen, Mike Canter, standing there looking at the people working, but I noticed that his eyes were very sad. He was slumped over a workbench.

"Good morning, Mike." I said. "It can't be all that bad!"

"Oh, yes, it is, Joe." Mike replied. "My wife has been in a hospital in Austin for twenty-two months as a result of a stroke she had while visiting her family there. She's in a coma and our children are crying for their mother. I have my mother here helping out, but we're running out of money!"

I told him I was sorry and hated myself for asking him. He had enough troubles without my butting in. I thought about Mike's problems and finally talked it over with my planner-engineer buddies. We decided to take up a collection for Mike and his family. Word soon spread all over the shops, not just electronics, but all the other shops as well.

The money was a real blessing for Mike and his family. His wife was brought to San Antonio and soon began to recover.

The next time I saw him in the radio shop, I just said, "Good morning, Mike."

"My family is home, Joe!" Mike said. "And I want to tell you it was nice what you started, my friend! Thank you!"

"No thanks are necessary." I replied. "I was just doing my job!" Mike was one of my best foremen in the production of black boxes.

LaRue was still teaching and we had one daughter in college at San Antonio College and the other attending Robert E. Lee High School. We had two cars and it was time to teach the girls how to drive so they could get their drivers licenses. LaRue took on the job of teaching them, Barbara first and Ellen later.

One Christmas Eve, we were enjoying a wonderful evening at home with good food, gifts, etc. Ellen's boyfriend, Hubert Nelson, was there, whom we knew quite well. We were friends with his parents from the old Harlandale Baptist Church days. Hubert would be graduating from Texas A & M University in June.

After everyone had gone to bed except Ellen, Hubert, and I, he asked if he could talk to me. He said he had a question for me.

"Ellen and I love each other very much and with your permission and blessings, we would like to marry in June. Do you have any objections?"

"Yes, I do!" I said.

Hubert looked shocked. "Why? Why?" He asked.

"You have my approval to marry in June, but only if you let Ellen finish college. She only has two more years to go at St. Mary's University. I'll pay her tuition, lab fees, book, and so on, but you have to take care of board and food." I smiled at him. "Of course, we'll help if you need assistance!"

"Why are you so concerned with her education?" Hubert asked. "I have a good job waiting and I can take care of her!"

"Well," I said, "What if you have children and you have a heart attack or some other illness? What would Ellen do? She's only worked as a checkout clerk in a grocery store. That wouldn't take care of her and the children."

"You're right." Hubert said. "I think it's a great and fair idea!"

We shook hands and that was that. It all worked out nicely. Ellen graduated and became a schoolteacher. We paid the college bills and helped out with incidentals.

Hubert has been one of my best friends and fishing buddies over the years. We've never had a quarrel or disagreement of any kind. He helped us with the mowing and repairs at both the shore house and our home. He's always been a wonderful son-in-law—more like a son to me.

His parents and LaRue and I fished together many times and attended the same church until they retired and moved back to their hometown of Leesburg, Texas, east of Dallas. Even then, we visited back and forth until LaRue became ill with cancer. They were a wonderful couple.

In spite of our very busy schedules, LaRue, Barbara, and I managed to attend as least two Dallas Cowboys football games each year as well as many Texas A & M games played in College Station, Texas. We all enjoyed those games very much, but our number one favorite sport was fishing.

Eventually the Department of the Air Force decided to see if would be more efficient to run it more like a business. There was too much duplication of responsibilities and too many overhead personnel at each air base.

Ernest and Ernest Co., efficiency experts, were hired to study Norton AFB in California, in spite of all the objections of the Chiefs who were entrenched in their little empires.

The outcome was that there would be no more deputy chiefs, all of whom had nonproductive alternates requiring clerks and typists to furnish them with the information they needed in order to function.

Under the new system, if the division chief were absent, one of his branch chiefs would sit in for him. The branch chief's alternate would be one of his sections chiefs, and so forth.

Rotation of acting chiefs would train people in the same division to handle their next promotion. He would understand better what was needed to support the program from his new position.

The program was implemented at Kelly Air Force Depot and it was unbelievable the amount of savings we incurred immediately. The program eliminated a lot of alternate officers, secretaries, file clerks, telephones, and, most of all, office space that was later used for production areas.

No one was fired. People were put into more useful hands-on production after their training. And, of course, all kinds of time-consuming weekly meetings held by every chief were eliminated. Instead there was one weekly or monthly meeting. That in itself gave the employees more time to do their own jobs instead of wasting time digging up all kinds of information to pacify the chiefs.

This was the best-case scenario of "Parkinson's Law of Diminishing Returns". Parkinson was a lieutenant in the British Army during WWI, serving on the battlefront in France. He discovered that as each commander above him was killed in action, the fewer reports he had to make, which required his own troops to give less information to him.

As the war progressed, he had to make very few reports. He evolved the "Parkinson's Theory": The more people at the top of an organization, the less time the employees at the bottom of that organization would have to do their own jobs! He was absolutely correct!

Norton Air Base went overboard and eliminated most of the radio/radar technician positions and hired 300 college graduates, electronic engineers, to do their jobs. After a few months, they had to recall all the technicians! They had all the good old "on the job" repair experience and, as we old electronics men would say, they knew where to look for the trouble each black box had, with its own peculiar problems in each system.

To improve efficiency and production, a three-month course for planners in engineering was implemented at Kelly Field's school building. The teaching staff was from Ann Arbor's University of Michigan. We had to be either licensed or certified.

The course was industrial engineering, or time measurement. We would work in our offices the first two hours each day and spend the rest of the day in school. There was a lot of homework to be done to survive. Only a few people from each unit attended the school because production had to be maintained.

It was a condensed course covering time studies of technicians and office workers and involved judging people working at a comfortable pace. I was fortunate to be able to judge workers in the training film exams correctly time after time, much to the surprise of the instructors.

Random sampling of a production line and being able to find problems that hindered production was my favorite thing. I used a simple method. I cut slips of paper and wrote times on them, starting with 7:30 A.M.; the next would be 9:00 A.M., and so forth for the entire shift of whatever unit I was going to sample. Then I put the slips in a hat and shook it up. I would draw out eight slips for Monday and record them on a chart. Then I would do Tuesday and so on the same way for twenty working days.

My first sampling was of a gas turbine engine unit that was consistently the lowest in production each month. I took a large tablet and each of the eight slips to the shop. All the workbenches were numbered, as was the foreman's office. I used codes—"W" for workers on the benches, "A" for absent, "O" for in the office, "T" for on the telephone, etc. After ten working days, I had my large chart completed and showed it to my supervisor. He called a staff meeting with the foreman of the unit and his supervisor.

I had found that the workers on benches seven and twelve were not working full time at their workbenches. In fact, they were in the foreman's office several times a day during working hours.

The foreman was stunned. "But they're my best workers! I don't understand!"

He was shown the chart on which I had recorded the absences of those two workers, along with others. I also pointed out that a couple of the other men had reading material and spent a lot of time in the restroom. Another made several telephone calls each day.

Corrective action was taken immediately and a month later this unit was near the top in production of engine units!

I was shocked to see that in our own office, we had employees who came in late, then called home to wake the family or let their spouses know they had arrived safely or just to talk. Some of the workers still had to comb their hair, apply their makeup, or simply chat with one another before they were actually ready to work. Men discussed their luck" or cursing the lack of it in fishing or on the golf course the past weekend!

A stop was put to that quickly. After all, we were supposed to be setting an example! The old adage "when you point your finger at someone, you have three pointing back at you" was so true!

After years of debate and planning, the Air Force began to consolidate repairs of large jet engines, starter jet engines, and accessories at Kelly Field. Electronics were sent to Tinker Field in Oklahoma, and other equipment went to Sacramento Depot in California. There had been too much duplication of repair facilities. It was a sad day for me to see Electronics leave Kelly Field.

After three months of training to be Industrial Engineering Technicians, we took the certification exam that was given by the representatives from Ann Arbor.

The pressure was unbelievable. If we didn't pass, we would not be given our next grade, which was GS-11. We were all nervous, but I had been similar spots many times before and soon calmed down.

It was quite an exam—carefully monitored and no distractions were allowed. I was finished at noon and turned in my booklet. I waited patiently while the lead instructor scanned my answers carefully. I had made a grade of 94%, but had to wait until it was officially verified at the Ann Arbor offices of the University of Michigan.

I almost let out a yell of joy, but said simply, "Thank you." I was able to contain myself until I got outside the building!

This was a Friday and my family had a four-day weekend off from school. They had driven down to the cottage in Rockport on Thursday morning. I had packed up our second car, a VW Beetle, the night before so I could leave for the cottage right from work.

I was so excited and relieved and happy that I had made it that the car didn't seem to go fast enough for the entire two-hour trip! Finally, I got a sudden feeling of warmth inside me and settled down, obeying the speed laws the rest of the way!

My family was waiting and greeted me exuberantly, even before I could tell them that I had passed the exam. They just knew by the way I looked that I had made it. I took them out to dinner that evening to celebrate!

On Monday, we did a lot of closing records and updating our labor standards for each type of black box, which went into the hundreds! I learned how to take a technical order on each one and come up with a revision of the previous labor standards. By comparing that with the actual hours it took in the previous year, we had an accurate labor standard.

I owe a great deal to a fine coworker who had been in Labor Standards for a couple of years and who was anxious for us to update all the records that were to be shipped to the other repair air base.

The movers came in one day with a floor plan of Building #308, our building, which had over 1,500 employees. They also had the floor plan for Building #323, a smaller building that housed Engineering on the second floor.

They marked our desks and prepared everything for the move over the weekend, which they carried out on time.

At our last Electronic Planning-Engineering meeting, I told everyone not to aggravate our new "buddies" in Building 323 by saying, "We didn't do it this way in 308", or anything else to upset them. I told them to just be quiet and see what happens but to do everything they could to help.

I had known the security officer in that building for several years and we had become friends. He clued me in as to what to expect there. I was glad that my tenure and additional duty as Chief Security Officer for Engineering-Planning for over twenty-two years was finally over. No more four safes with four drawers filled with classified documents to check every evening. Those four sets of combinations were difficult to remember and we weren't allowed to write them down, although some did. They would write them on their desktop daily calendars, with the three numbers scattered on different pages. But this was the first place the inspectors looked when they did security checks!

We were formally greeted in Building 323 but in a reserved fashion. I was to supervise seven men who were not from Electronics. They were from Small Jet Engine Engineering.

We were immediately busy and proceeded to overhaul the time labor standards on each type of engine as well as the accessories. Gradually, the "ice melted" and we became a team.

My responsibilities increased. I was given the task of overseeing the use of forty microfiche scanners and eighty-four computer/tape displayers, which had technical orders on videotape and could make an instant copy of any page needed. These were located in shops in different buildings.

I did the overhead budget, which was for non-production employees such as supervisors, clerks, secretaries, production control people, and the scheduling and quality control people. Then it had to be entered into the computers that were online with Headquarters at Wright-Patterson in Dayton, Ohio. This had to be done a couple times a year. We were allotted 184 lines in the computer for eight sections of our branch.

I know I lost four or five pounds in sweat each time I spent four hours deleting the old figures and entering the new ones. Great care had to be taken not to use any incorrect letters in closing the computer. When the computer screen read "good night", one simple error could erase everyone's budget!

My repairmen were scattered in four different buildings, the farthest was a half a mile away, across the concrete ramps from my office.

One day, as I was walking to the Headquarters building farthest away to check on production on the hanger floor, I passed an old Mexican-American friend. We shook hands and he said, "Joe, I'm on the Equal Opportunity Committee with eight other people. We take all the information on anyone who is up for promotion and make sure each one is treated fairly and that a qualified minority person is not being bypassed for promotion. Your name came up for a GS-11 and we all recommended to the Chief of Personnel that you be promoted!"

He looked at me. "Now, I'm not supposed to tell you this, but I just had to! I'm so happy for you!"

I was quite shocked that all this was going on but I thanked him for his support and information.

It was the custom that when five or six people received a promotion, they would throw a "beer bust" for their fellow workers and supervisors, although some didn't do it.

We had ours at a park on part of the nine-hole golf course at Kelly Field. It was to be held at 4:00 p.m., after work. Our friendly brewer delivered two kegs of beer, complete with pumps and ice cold. Then a fast trip to the tamale factory to get two five-gallon cans of tamales, 144 tamales to a can, plus paper plats and plastic cups for the beer.

Needless to say, the party was a huge success! Everyone shook hands with the six of us who had been promoted. I didn't know I had so many friends on the base! We had also invited some of our closer buddies we had worked with in other branches.

It was party time at home too when I received my first paycheck! What a difference that promotion made in my pay!

The biggest difference was at Kelly Field. There were a lot of "perks" for supervisors and I enjoyed them all!

My desk was moved once more to a newly constructed inner office at ground level in a the building where the gas turbine engines were disassembled, checked for flaws, cleaned, and reassembled with whatever new parts were necessary.

I had six Engineer-Planners, a secretary, and a private office! I never had a more supportive group of men anywhere than this one. In a year's time, we had our production up to 96%, way above what they had before. I was kept busy running from building to building to check out production problems, material arriving late, and breakdowns. It was a lot of pressure.

My supervisor was quite pleased and decided to swap me for the supervisor of another section, similar to mine, in their own building where their offices were located. The offices were separated by walls and a door!

Henry Valdez had only gotten 86% production so I had a challenge to bring it up. I began to see the problem immediately: the "chiefs" were too close to my "Indians"!

At the very beginning, I had a meeting with my men. I told them, "If anyone asks you to do something or go anywhere, tell them they have to see or call me first!"

After a couple of weeks, one supervisor said to me, "Diviak, you run a tight rein on your employees. One of them wouldn't even help me bring in some office equipment unless I asked you first!"

"Chief," I replied, "That is only one of the reasons this section ran only 86% effective in production last year! There are millwrights only a phone call away to do the moving and lifting! That way we office people don't 'herniate' ourselves. They have the equipment."

We dug into all the labor standards and parts requirements for these particular jet engines to be repaired and within a year, we were 96% effective. It would have been an even higher percentage but the parts needed for one type of engine were not available and we had to get them from an alternate source. We repaired hundreds of small jet engines a year.

Every summer, we had "summer hires", high school kids who worked on a temporary basis. We hired them not only to type and file, but to show them what they could do if they tried and stayed in high school. They were mostly minority children and they learned a lot.

One little lady, Henrietta, came to work in baggy slacks, sandals, bushy hair and a sleeveless blouse. She worked in the outer office. At lunchtime, I generally had my sandwich and coffee at my desk. One day, Henrietta asked if she could talk to me.

She had her own "brown bag" and we kept a coffee urn going in the outer office. I asked her to sit down and she began to talk about her home life and frustration at getting summer jobs. She asked for my advice on how she could do better.

I told her not to squander her first paycheck but to buy some nice jeans, blouses, and closed-toe shoes. I told her, above all, to ask a well-dressed friend for advice on hairstyles and makeup. I told her that in any office, everyone should look neat and present a pleasing picture to the people doing business with that place.

The next morning, Henrietta had made some changes to her appearance. "Beautiful!" I told her.

The Monday after payday, Henrietta had on a nice outfit and I hardly recognized her! She was a beauty! Everyone came up and greeted her warmly. During the rest of that summer, she became the most well liked person in the main outer office. Her typing and filing was the best.

She received a nice letter of recommendation from my Chief when she left that fall to return to school.

She came into my office all dressed up in high heels and all to say good-bye. She looked like a beautiful, grown-up lady. She came up to me and kissed my cheek and gave me a hug. "I wish I had a father like you! What a difference it would have made. You really helped me. Thank you! Thank you!" I never saw Henrietta again, but I'll bet she did well.

Time seemed to fly by and suddenly I was checking out retirement to see if we could live on my retirement pay and LaRue's teacher's retirement pension if we retired.

I was also fearful that I wouldn't have anything to do in retirement so I enrolled in night classes in real estate sales at San Antonio College. I could take the state exam but first I had to find a realtor whose company would sponsor me. A Benevedes realtor agreed to do it and I enrolled for the Tuesday night classes. I loved the course, which turned out to be three hours a night, two nights a week. I'd already had a Commercial Law course in high school and a year in Business College in Roanoke.

At the end of the course, we took the exam given by the college. All who passed that could then take the state license exam. I passed both with a 92 on each!

I worked weekends at the real estate office. I made a lot of phone calls but never received credit for any of them nor any money even though I sold three pieces of property. I finally eased up on that because we were planning to move to Austin Lake Shores near Austin, Texas when we retired.

Being an avid and a good fisherman, my lifelong hobby, I figured I could at last fully enjoy it. We still had the beach cottage in Rockport, but to live there permanently, the hurricanes and salt air wouldn't be too good for the nice furnishings in our home or for our cars.

I told my supervisor that I was planning to retire in a year, but he just laughed. I finally got him to let me train a couple of alternate coworkers on the additional duties and the overhead budget. No one would believe that I was really going to retire but I did! It was so quick that they didn't have time to plan a big party in town so they held one in the office where I worked.

Before my retirement day, I had to attend a couple of debriefing classes, given by doctors, nurses, and businessmen. It was a good class but the only thing I remember is one lady nurse, who said, "A lot of retirees start drinking due to boredom. That's a terrible thing to do after you've worked so hard all these years so you could enjoy retirement!"

A tiny voice from the back of the room called out, "I'll drink to that!"

That broke up the meeting!

CHAPTER NINETEEN

Federal Civil Service Retirement

I "walked my clearance through" all the tool cribs, library, finance department, etc., to be signed so I could secure my retirement papers, scroll, and awards. There were no problems. I got a lot of handshakes, kisses, hugs, and a few tears, too.

I left Kelly Field on July 4, 1976, at age sixty-one.

My supervisor was almost bitter when I left. As I drove through the security checkpoint at the gate and turned in my identification badge, I recalled that I had landed at Kelly Field for the first time in November 1937. We had flown from Chanute Field in Illinois in a Dutch-made U.S. Air Corps 5-passenger single engine aircraft with fixed landing gear. The cabin was nice but the pilot was outside at the back of the cabin in an open cockpit with only a windshield to protect him!

I drove home to my wife, LaRue, who had retired in June after twenty-six years of teaching.

LaRue C. Diviak, retired in June, 1976, after twenty-six years of teaching school.

We were married nearly forty-five years when she died in May, 1985

A wonderful lady, mother, wife, pal and friend, who was loved by all who met her.

It was wonderful to be able to rest for a few days then drive to Austin to visit her brother and sister-in-law. Don was a partner in the Bradfield-Cummins Realty Company in Austin. I had moved my real estate license to Don's company to keep it active and current in case I needed it.

Don and Betty Ann had a large beautiful home on Lake Travis. We began looking for lots for sale and finally bought a large one in a new development of 1,000 acres. Our lot was on a hillside looking over the upper end of Lake Austin, about a mile from Don's house across a pass in the hills. There are five lakes, one above the other, on the Colorado River. There was plenty of room to fish and just enjoy the scenery.

We had a three bedroom, two bath home built in a factory in Austin and trucked out to our lot. They set the home on eight-foot concrete pillars firmly anchored. Later, I built covered front and back porches the length of the house, about forty-four feet. The back porch was twelve feet off the ground and I built a sturdy rail so no one would fall off.

We had a fantastic view of upper Lake Austin and the rolling hills around it.

We found a quarry about twenty miles from us that sold cut rocks that were rejects for $2.00 a truck or trailer load. I had a truck and a trailer and LaRue and I hauled loads of rocks to our property. We built retaining walls and floored the basement and enclosed it. In all, we used forty-two tons of rock! In front, we built a two-car garage.

LaRue was a great helper and was extremely happy in our nice large home in a quietly scenic neighborhood.

The Sub-division was called "Apache Shores" and there was a lot of "perks" for the residents. We had our own picnic area on the lakeshore under live oak trees, and piers. There was a concrete boat launch ramp and parking spaces for the vehicles and boat trailers. Up at the entrance was a large community hall, with a dance floor and a kitchen for parties.

And, best of all, at the gate was the emergency fire and ambulance station. There were forty people trained in CPR and such, all volunteers. We were only eighteen miles from Austin. It was a fantastic place to live. Even if we lived 100 yards apart, we were all neighbors and helped each other.

When I was a teenager, I saw a movie where there was a cottage on a lake. Even in black and white, it looked fantastic. Since then, I had dreamed of living on a lake or shore. Over the years, that dream never diminished and now I had that house! A wonderful dream come true!

The only negative point was that the Mansfield Dam was up from us. The water coming from the bottom of the lake through the power generating turbines was sixty degrees or colder all year. It wasn't conducive to swimming or skiing.

In the early days of Lake Austin, there was a duck-hunting club in Austin. They hired an ex-Texas University horticulturist to plant duck weed on the banks of the lake to induce ducks to land their on their flight south to the Gulf coast.

The results were devastating. He planted the wrong thing and it spread all over the banks, below the water level, and in the shallow water all over the upper level all over the upper half of the lake. Worst of all, the ducks didn't like the stuff!

You couldn't swim in the tangled vines and boat propellers were always getting jammed. The roots were below the water level, as far as eight feet down. You couldn't even fish in those vines without losing lures, line, sinkers, and such.

Now, every two years in February, during the coldest week of the year, the gates at the dam were opened enough to cover the lake eight to ten feet, to allow the "duck weed" to freeze.

We never seem to learn to leave nature alone! A good example is the water lily brought in bowls to the World's Fair in San Francisco years ago, and also the water hyacinth. Both choke most of the lakes in the south. Oh yes, and the beautiful vine that was brought to the southern states that now costs millions to control its growth all over the trees and power lines!

The time came when LaRue and I were living alone. Ellen, our oldest daughter, was a schoolteacher in Bryan, Texas and her husband, Hubert, was a city planner. Their two children were now in school. Barbara, our youngest, was teaching school in San Antonio.

LaRue and I began driving to the coast to fish and sightsee, do some beach combing, and relax in our cottage on the shore at the Bay. We took trips to Colorado in our truck to visit friends. It was fun.

I became a Sunday school teacher to a small country Texas Southern Convention Baptist Church, in Spicewood, to a class of thirteen or fourteen men. LaRue played the piano and her good friend, Marge Stoup, was the church organist. The men in my class were all retirees from Federal Civil Service and ranchers.

We had a monthly business meeting and Church socials once a month. Each family usually brought a "covered dish" to the socials. The food was always fantastic and so was the caring fellowship.

CHAPTER TWENTY

Medical Problems and the Loss of LaRue

One beautiful June day, we drove to our cottage on the coast to spend a couple of days fishing and visiting with neighbors and friends in town.

On the morning of the second day, LaRue awoke with a fever. She also had a spot on the sole of her left foot that looked like a blister about the size of a quarter. I immediately called a doctor whom we had seen there before and she came out to the cottage to check LaRue and see what the problem was. LaRue had been bitten by a brown recluse spider, also known as a fiddle back spider. They hide in bedding and cracks inside and outside houses.

The bites result in flu-like symptoms and can be fatal if not treated quickly. The doctor gave LaRue a shot and some antibiotics to take. After three days, the fever was gone but it took weeks for her foot to heal completely.

Then LaRue's weight seemed to balloon. She had to be treated for excess fluid in her body. A year later, she felt a lump in her left breast. We went to a clinic to have it checked. It was cancer in the early stages. She had surgery and tests indicated that it was the least deadly of six or seven types of breast cancer known at that time.

LaRue underwent thirty cobalt treatments and soon recovered enough to travel a little to West Texas, Colorado, and West Virginia.

In the meantime, our youngest daughter had met and become engaged to a young man, Dave, who was a carpenter and building contractor. They were married secretly in Austin and moved to Apache Shores. Dave turned out to be an alcoholic and a real pain and a few months later, they separated.

A year after her marriage, Barbara had a beautiful baby girl, Stephanie, who was born with a heart defect. The baby would need open-heart surgery to reverse two of the arteries. When she was three months old, the surgery was performed in the children's section of Santa Rosa Hospital in San Antonio. Stephanie died after the surgery and was buried next to our daughter, Helen, in Bryan Cemetery.

Dave came to the funeral and later, to our home to say goodbye. I told him that he was no longer welcome in our house after that. We had spent a good deal of money getting him out of trouble. All LaRue and I wanted was to enjoy our retirement with what little time we had left.

Dave cried and said, "Joe, you and LaRue are the nicest people I've ever met. I'm really sorry I can't stop drinking, but I thank you!" We never saw him again.

Barbara divorced Dave and regained her maiden name. She went back to teaching.

In Texas, there is a community property law and spouses are responsible for half the debts on credit cards made by their spouses. It took my daughter years to pay off her half of debts Dave had made at various places.

LaRue and I finally had room and time to breathe the nice clean air of the hill country and go for walks in our Apache Shores area. We made three trips to Colorado to visit friends there. The fishing was terrible! Every creek and river was overrun by senior citizens fishing, stumbling around, and shouting at each other. Others were sitting in deck chairs on the banks of lakes and ponds, their fishing rods stuck in holders on their chairs, reading newspapers and books.

We had to wait in line for gas and even at restaurants. I vowed then that I would never go back there and I didn't after that last trip in 1972!

And, oh yes, the streams were full of algae and scum and I couldn't figure out why. Finally, I looked around and saw that there was a sewage holding pond downhill from each lovely home. Of course, every heavy rain would wash the water and sewage into the streams below.

We sold our Rockport beach cottage for three times what we had paid for it and used the money to pay off our home on Lake Austin. We were finally debt-free!

I became involved with the school board and the general elections for two precincts in Travis County. I was asked to be election judge by our lovely county commissioner, Ann Richards, and served for four or five years. I met a lot of interesting people who lived in a wonderful area called Key Allegro, mostly retired people.

Our family doctor had his offices there. He and his wife were wonderful people. Dr. Haule was an ex-Air Force flight surgeon and had built a terrific practice in that area. He was the one who did LaRue's breast cancer surgery.

After about five years, LaRue began to feel pain in her pelvic bones. X-rays showed tumors growing on the bone. High intensity X-rays took care of the tumors but cancer reappeared six or seven months later, resulting in more hospitalization and more radiation treatments.

This went on for four years. The last six months, she was terminal and I took care of her with the help of a visiting nurse who came in for a couple of hours twice a week.

I learned how to give LaRue morphine shots in her thighs every four hours. She would cry and ask me to "accidentally" give her an overdose. She wanted to go peacefully with no more pain. I told her I couldn't do that because if the coroner check and found out I had done that, I would be arrested for murder.

LaRue finally had to be admitted to St. David's Hospital in Austin when it was found that the cancer had spread to her liver. She passed away a week later in her sleep.

About four months before, she had begged me not to sit around and mourn her death, but to go out and enjoy life and have fun. "Joe," she said, "You're from a family of long lived people. Don't waste the years by being true to me. When I'm gone, that's it. I want you to be happy. You've been a wonderful husband and father, so enjoy life! If you meet someone you can love, go ahead and marry her!"

That upset me at first but I got to thinking that if the situation were reversed, I would want her to go on and live her life to the fullest.

The funeral was held in Bryan, a two-hour drive from Austin. LaRue was buried near our daughter and granddaughter in the family plot. Our young pastor, Reverend Don Dickerson, conducted the service in the chapel at the Bryan Funeral Home. We sang LaRue's favorite song, "Amazing Grace". LaRue had requested that the service both at the funeral home and the cemetery be short.

CHAPTER TWENTY-ONE

After Fifty Years, A Letter From My Childhood Sweetheart

About five months before LaRue died, I received a letter that really surprised me, especially when I saw the return addressee's name. It was from Mrs. Ruthie Sands at an address in Philadelphia, not Woodlawn, VA.

I knew immediately it was from that sweet girl I had met when we were sixteen years old, back in Caretta, West Virginia. I sat down on the steps by the mailbox. I felt a little uneasy for a moment, then I opened the letter.

Ruthie had sent the letter to my sister, Theresa, who lived in War, West Virginia, asking that she forward it to me. Theresa sent it to me with a note saying that she had written to Ruthie telling her she would be hearing from me.

It was a nice letter, all about her life after we stopped writing to each other. She had gone to nursing school, married and had a daughter, Jackie. She was later divorced. She had worked for a doctor in private practice in Baltimore, Maryland, as a registered medical assistant, for twenty-three years. Then she got into real estate for a firm there and became a very successful saleslady, and vice president of the largest residential realty company in Baltimore.

She had eventually remarried, moved to Hillsville, Virginia, and her husband had died there after a few years. She wrote that she had been going through her trunk of "precious possessions" and came across a Whitman's candy box filled with my love letters to her. The box still had the same rubber bands around it that had held it shut for almost fifty years.

She wrote that she had started to read the letters and then wondered what had become of me. She got the idea of trying to locate me and remembered that my sister lived in War, near the high school I had attended.

She said that if we lived anywhere near Philadelphia, to come visit her. Just let her know ahead of time so she could make some plans. The cutest thing about her letter was the way she started it: "Dear Joe and your lovely wife, I know you must be married—you're too nice a man not to be!"

I took the letter into the bedroom and read it to LaRue. She smiled as I read it and when I was finished, she said, "It's a wonderful letter! I want you to answer it as soon as possible. I remember you telling me that she was the only one you ever had a date with before you met me!" That was true!

Within a week, I answered Ruthie's letter, giving her a brief summary of my life since we'd last communicated. I had LaRue read it and she said, "It's perfect, Joe!"

LaRue knew she didn't have long to live and was welcoming a quiet and peaceful passing. Ruthie wrote to both of us in every letter and sent LaRue "get well" cards and notes of encouragement and hope.

After LaRue died in May of 1985, I became so relieved because at last, LaRue was at peace. I still woke up in the middle of the night, ready to give her a morphine shot. Then it would occur to me that it was over and I wouldn't have to do that any more.

Neither of my daughters could help me much during LaRue's long illness. As schoolteachers, they had to be on the job every day. LaRue was my patient and I wouldn't have wanted it any other way.

A few days after the funeral, I went to the two hospitals, the radiology departments and the doctors and nurses who had helped the most and thanked them all. They were surprised because, evidently, that didn't happen too often. We shook hands and hugged and wished each other well.

One really cute nurse came over to me and said, "Mr. Diviak, one of these Saturday nights when you don't have anything to do, call me and we'll go 'bar hopping' and dancing! Nothing serious—just have some fun!" I thanked her but I never did call her. But I knew she had the right idea: unwind and live again.

Nowhere is it written that you have to mourn and wail for a given length of time. The quicker you get on with your life, the better. I had a couple of calls after that, but I never went out on a date, although I felt flattered that they cared enough to ask.

The Diviak Family Reunion was held in Arlington, Virginia, at my brother's home. He was a retired coalminer and lived on a couple of acres near a lake. It was a beautiful spot and a great place with space for tents and RV's. The reunion was held on the third weekend in July 1983. About eighty-five people showed up. My daughter, Ellen, had gone with me. We drove there from Texas in my Ford super cab long wheelbase pickup, which had a self-contained camper on it. The camper slept four so we had a lot of room and had everything we needed to cook, sleep, etc. We parked in a campground a couple of miles from brother John's home.

It was a lot of fun. Ellen wrote down all the names of those attending the reunion and the history of the meeting.

That July after LaRue's death, the reunion was being held in Bumpus, Virginia, near Richmond, at my sister, Helen's daughter's home, where Helen had lived since her husband had passed away.

I wrote to Ruthie about it and she replied that she always went to her sister Mae's home in Hillsville, Virginia for a month, but she could move it up a couple of weeks and that I could meet her there. Mae was also a widow.

I drove there in my "Texas Cadillac"—the Ford super cab pickup, this time without the camper. It was the best vehicle I'd ever driven. It had a radio and tape player, and the seats felt like upholstered chairs. The driving was easy and I had a good view over the cars ahead on the highway.

When the time came for me to leave for Virginia, I was a bit apprehensive and it seemed to build that first day of the trip. I stopped at a motel in Jackson, Tennessee, had a great dinner and a good night's rest. In the morning, I enjoyed a real southern-style breakfast of biscuits and gravy, ham and eggs, and coffee.

I had gotten an early start for Hillsville and still I had that strange feeling, but I drove on. Finally, about two hours out of Hillsville, everything seemed to calm down inside me and I was happy again.

I arrived in Hillsville in the early afternoon and pulled into Mae's driveway. Out walked a slim, beautiful lady. I got out of the truck and walked toward Ruthie. We hugged then we kissed. After fifty years, the old feelings returned and I was really happy. She was a beauty and greeted me warmly. It has been said that you can only love one woman, but this love was different and both were wonderful in their own way.

We went inside and I met Mae, Ruthie's sister. I stayed there for a couple of days. We had a wonderful time together. Then it was time for me to continue on to the reunion. It was a two-day celebration with ninety of us there. Then I returned to Hillsville.

Ruthie Horton Goad, age 62, member of the Million Dollar Club. Retired in 1975 as a vice president of the largest residential real estate company in Baltimore County, Maryland, after having worked 23 years as a registered medical assistant and registered medical secretary to a Doctor of Internal Medicine and Cardiology in Baltimore, Maryland. Author: A Courageous Woman

Diviak Reunion, July, 1985 in Bumpas, Virginia
My seven siblings and me – Mary had died in 1884.
Front row: Julia, John, Steven and Josephine
Second row: Joseph, Helen, Anna, and Teresa

Ruthie wanted me to meet her daughter, Jackie, and her family in Philadelphia, so we drove to Philadelphia for a couple of days. I met all of Ruthie's friends in the Chestnut Hill apartment complex where Ruthie had a beautiful apartment. It was all like a dream.

Ruthie's daughter and her husband, "Hap", and their children, Laura and Ian were all so sweet.

Ruthie bowled in a ladies league so I had to learn to bowl since there were mixed leagues at the Center.

After a couple of days, we headed back to Hillsville, where I met her brother, Guy, and his wife, Ella, who lived next door to Mae. I had known Guy in my coalmine days in Caretta. He had lived near the company store and we had seen each other there often, as well as in the mines. He was a gas inspector, a person who carried a specially equipped lantern that showed the presence of methane gas in the mines. If gas collected in an area, it had to be corrected immediately by removing the obstructions or redirecting the airflow. It was a tough job and required a lot of bent-over walking through the four-foot high seams of coal.

All of Ruthie's relatives there were nice and friendly. The ones who had worked in the old Caretta Mines had known my father for years.

After I had almost overstayed my welcome, I left for Texas. It was truly hard to say goodbye.

I had invited Ruthie to come visit me and meet my daughters, and see how she liked Texas. My daughters and I got together and made a list of things they wanted and what would be sold since there wasn't room in their homes for all the stuff I wouldn't need, even if I stayed in Austin.

Two sales netted several hundred dollars. The clothes we didn't sell were taken to a church mission group in Bryan to dispense to the needy. All of LaRue's jewelry went to our daughters along with other items she had left.

After a while, I began to detect a tone of resentment about my being so happy and fun loving. I found out later that they had discussed the possibility of having me committed for a while, until I "returned to my senses"!

When I found out when Ruthie was coming to visit, they informed me that they were not ready to meet her! What a shock that was to me! I had this weird idea that if I was happy and they didn't have to take care of me, they would be happy for me and glad that I wasn't brooding and being difficult to cheer up.

I had helped rear my daughters and thought I knew them better than that. We had camped out, fished together, gone to football games and church together, had eaten all our meals together, had been a family!

When Ruthie arrived, I picked her up at the airport and took her to my home. I had a great dinner all prepared for her. She was a doll! I put her things into the guest room. She had her room and I had mine.

I took Ruthie to meet LaRue's bother Don and his wife Betty Ann. They greeted Ruthie like a long-lost friend and made her feel welcome.

Don was terrific and said that I had been a real husband to his sister and he would never forget how good a friend I had always been to him and his family over the years.

Ruthie and I went to a college football game in Waco. It was Baylor and my all-time favorite, Texas A&M playing. Then we went to a Dallas Cowboy's game when they played the Houston Astros.

Then we went to dinner on the Patio on the River Walk where I dined many time before over the years when we had lived in San Antonio.

When we returned to my house, Ruthie asked, "When do I meet your daughters?" I explained that they had said they "weren't ready yet" but that I would call and see what they said.

When I called each of them, I got a very negative response from both of them. That shocked Ruthie and I was completely stunned. We couldn't believe it!

At the time, I didn't know that they had problems of their own but still, they could have been more receptive, especially after all we had gone through together.

I knew then that Ruthie would never move to Texas and beside, she didn't like the hot weather. We were having a very hot and miserable summer and fall that year! Nor did she care for my ranch-style house with the vaulted ceilings on the lake. What I liked most about it was that it was paid for!

Ruthie had Chippendale and Queen Anne and other antique furniture and it wouldn't fit into that house.

All my neighbors in Apache Shores loved Ruthie immediately. We all lived a hundred yards or more apart—all the lots were two- to two and a half acres. Two of them invited us for dinner, one couple to their home and the other to a restaurant where the woman was the chef. Then a fellow church member and his wife invited us to his home, along with several other couples, to meet Ruthie. It was a dream come true, but still the lack of compassion from my daughters hung like a black cloud over my head.

Before Ruthie left for Philadelphia, we were enjoying some dip and cocktails in the living room. Ruthie was on the sofa and I knelt down on one knee on the carpet in front of her.

I told her that I loved her very much and asked if she would marry me. She hesitantly said "yes" but was concerned about my daughters' attitudes. She didn't want to cause any problems between them and me. I told her it would take time, but they would change.

I also felt that at our age, we couldn't afford to wait. We were both seventy-one years old and why not enjoy the years we had left!

In the end, we decided to wait a while to see if things would change.

I took Ruthie to the airport and she flew back to her home in Philadelphia. At the time, I still had hope that things would change.

All my friends seemed to have the same opinion. "As good as you've been to those two daughters and to their mother over the years, they should wish you a life of happiness! They sure can't take care of you if you have serious health problems, being they're both schoolteachers!"

After a while, I felt like climbing the walls! It was 8:30 in the morning. I got into my truck and went searching for a bowling alley. I remembered seeing one on IH-35 above Austin. Ranch Road 620 ran by Apache Shores and skirted Austin to IH-35, right near the bowling alley.

They were having league bowling but there was a lane open. I rented bowling shoes and selected a nice heavy ball with finger holes to fit my fingers. Then I threw a few balls. A man sitting nearby showed me how to keep score. After a while, I began to have some success and even threw a few strikes!

An eight member mixed league team started bowling near me. They were one "body" short. A nice lady asked me to join them. After making it clear to them that I had never bowled before, I did join them. That was fine with them. "They couldn't bowl either," the lady said.

During the game, they asked me if my wife bowled. I told them that she had died and that I was presently living alone in Apache Shores. Their eyes lit up. "That's only across Lake Travis from us!" they said.

We won three out of four games. I don't recall what my scores were but they were happy to have me!

Before we left, they invited me to a barbecue they were having that evening at around six o'clock. Someone drew me a map of how to get to the place. I went with a bottle of wine and had a good time. The music, drinks, barbecued steaks, and all the other food was just wonderful. I kept thinking how much Ruthie would have enjoyed this party.

As I was leaving with another guest, one of the ladies came over with her husband and invited me to another cookout the next night, just a short distance from where we were then.

The following night, the same group was at the cookout. This time, though, the food was Mexican and the drinks were margaritas. Once again, it was a fantastic evening.

The following week, I made reservations for Friday evening at 7:00 P.M., for five of us, at a family-style restaurant just off Ranch Road 620. I called my new friends and they said they would be happy to be my guests. They hadn't been there before.

We had our dinner in a very nice private room in the restaurant. The main dish was chicken-fried steak with a choice of rice pilaf or mashed potatoes, and corn on the cob, pinto beans and corn bread—all southern-style. After dinner we were served coffee and pecan pie with ice cream.

It was quite an enjoyable evening and we had a hard time saying goodbye that night.

I never saw them again. I had so many things to wrap up before moving to Philadelphia.

We did exchange Christmas cards and they always asked for me to bring my bride to Texas to visit them. That's the kind of people I miss most, because there are so few of them around.

During all this time, Ruthie and I wrote sweet letters to each other and exchanged poems and tapes. I think the one that surprised her most was my singing a song I made up using the tune from Dean Martin's The Good Times. At the end of the song, I played the tune on my harmonica. In my younger days, I played harmonica and accordion "by ear".

Her poems were so sweet. One I cherished the most was called "My Yosef". We constantly exchanged gifts and cards that meant a great deal to both of us.

Ruthie called me one day and asked if I would fly up for Christmas. I booked a round trip ticket to Philadelphia at a reduced price. My brother-in-law, Don, took me to the airport in Austin.

I was sitting near the boarding gate when a beautiful lady in a blue uniform sat beside me with all her carry-on luggage. We began talking and she asked where I was going.

I told her. She thought that was very nice. She kept asking questions while we waited. I noticed that her nametag above a set of wings said "Mary Moore". Finally, the plane boarded and we headed to Houston to pick up more passengers.

Once we left Houston and were up to cruising altitude, the stewardesses began serving our lunch. I ordered a small bottle of wine to accompany the food. The stewardess returned with a large bottle and would not accept my money! She told me the wine was "compliments of the crew".

Soon she returned with a lovely red rose in a bud vase and sat it on my tray. "This is from our pilot, Mary Moore." She said.

"We've got a lady flying this plane?" I asked.

The stewardess laughed. "Yes! Haven't you noticed how much smoother this flight is than usual?" The other passengers around me joined in my laughter. What a great flight! I felt wonderful.

As we were leaving the plane, the pilot's compartment was open and Mary Moore stood in the doorway with a huge smile and a warm handshake. I gave her a big hug and thanked her. Earlier, she had sent a second vase with a rose for Ruthie. I told her I would never forget her and thanked her again.

Ruthie met me at the terminal and when I explained the roses, she was very pleased. Later, when I returned to Texas, I wrote to the personnel section of the airline, telling them what a fine lady their pilot, Mary Moore, was and how warmly she treated the passengers and me in particular.

I received a reply from them saying that they were well aware of it and that a copy of my letter would be placed in her permanent file to be used in her future promotions.

Ruthie and I had a wonderful Christmas with her family and friends at the apartment complex. Family friends greeted me as if I were their own. But finally the time came for me to return to Texas. Don picked me up at the airport and delivered me to my home.

There were many things I still had to do, such as give my daughters and granddaughter whatever furniture they wanted, along with the things LaRue had wanted them to have. Ruthie told me to sell the furniture since there was no room for anything more in the two-bedroom apartment.

By April, I had sold most of the furniture. It was waiting to be picked up, having been left there for my use until I moved. We had garage sales and the boat and trailer was sold, along with the truck camper, the TV, band saw, router, and other tools, plus the garden tractor with the plow and mower, Weed Whacker, etc.

By June, all that was left was the bedroom furniture, some dishes, and a few other things that I would take in my covered pickup. Then I purchased a seven-foot high, drop-leaf desk with four shelves and three drawers, a seven-foot high bookcase, and a chest of drawers, all of solid mahogany, that I would need in Philadelphia.

In June, I went to see Dr. Haule for a checkup. He noticed that a mole below my right shoulder had turned a strange color.

Several months earlier, I was building a carport with a sundeck on top at my youngest daughter's home near my home in Apache Shores. I put a nice white railing around the deck and built steps leading down to the ground and installed outdoor carpeting on the deck. While I was doing that, I had somehow scraped the mole off. I put some antibiotic salve on it and it appeared to heal just fine.

Dr. Haule deadened the area around the mole as I lay face down on the examining table. Then he cut out the mole and a plug of the surrounding area about the size of a nickel and over a half-inch deep. He showed it to me then placed it into a container to be sent to the lab for analysis. He sutured the wound, bandaged it, and gave me a prescription for an antibiotic.

A couple of days later, he called to tell me it was a melanoma and wondered if he should remove more of the surrounding area. I told him no, that I thought it would be fine.

When I was ready to leave Austin, he gave me my medical records to take with me to Philadelphia and told me to have a doctor check the area every six months, which I did. I've never had any more problems with it.

I sold the house finally, after doing a bit more repairs and having it checked for termites, etc. I was finally ready. Ruthie flew down to help me pack my things and to travel with me. I had said my goodbyes to my family and friends. It was time to move on.

We drove to Hillsville, Virginia, and visited Mae, Ruthie's sister, on our way to the Chestnut Hill Apartments in Philadelphia.

We were married at a Holiday Inn's private chapel and had a wedding dinner a private dining room, with a musician playing a keyboard. It was a wonderful affair. There were over forty guests there and they all seemed to enjoy it.

Don Cummins, LaRue's brother, was my best man. His wife, Betty Ann, did most of the announcements and gave a nice speech before the toast and cutting of the cake.

Ruthie's grandson, Ian, was the ring bearer, and her granddaughter, Laura, was the flower girl. We had guests there from Maryland, people from the apartment complex and some from Texas.

Ruthie had sent wedding invitations to my daughters. One answered that she and her family couldn't make it. The other chose to just ignore it. But I was not going to let that ruin my life! I promptly shut it out of my mind. Life is too short to worry about what might have been and that becomes history.

Joseph T. Diviak and Ruthie H. Shands were married in Philadelphia, Pennsylvania, and live there now.

I don't believe my daughters wanted me to leave Texas. That could have been one of the reasons they opposed my marriage. I was always there when they needed me to repair something, build something, or do errands for them. I was "Mr. Fix-It" to them and I was very glad to do it.

I am a "master" of many trades–plumbing, electrical work, masonry and concrete work, tile setting, etc., etc., and made a living doing them in my lifetime.

During my nine years in the Army Air Corps, from Private to Master Sergeant, to Warrant Officer, all of which included WWII, I was trained to be a leader. That required me to be able to make quick, intelligent decisions and be right! The same was true of the twenty-seven years I spent in the Air Force's Civil Service. I was trained as a leader because of my background of doing things right the first time. I never had a problem solving the most difficult tasks.

When someone says, "You can't tell Joe anything because he'll do it his way anyway!" they're right! You can tell me, but I may not do what you told me if I think there's a better way! Experience is the best teacher and I've had more than my fair share of it. I was the person responsible and if I screw up—the army had a more descriptive phrase for that! —I'm the one who is going to be blamed, not the person who told me how to do something. They can always say, "Joe misunderstood what I said" and crawl out of the mess spanking clean!

Perhaps most children are afraid they will lose "their" inheritance, valuable items and, of course, the affection of their surviving parent, when that parent decides to remarry. They think the new person is trying to take Mom's or Dad's place and their deceased parent will be forgotten. No one can ever do that!

A better thought would be to realize that Mom or Dad would be gaining a meaningful, loving relationship with a mate to be with for the rest of their short remaining years. This new parent may turn out to be the best friend they ever had!

Then there are some children who want to keep the surviving parent with them. But after a while, they begin to complain that they have to stay at home to care for the parent, or that they can't accept invitations to social functions because their parent can't be left alone. You can receive a lot of self-serving attention and sympathy when you are "suffering" that way–sort of Munchausen Syndrome in reverse! I have seen it all!

I found out later that my daughters had a lot of problems of their own, which they didn't share with me or ask for my help, because they felt I'd had enough problems of my own for nine straight years with LaRue and her cancer before she died.

I won't go into their problems here. I forgave them long ago. They are my children.

A man couldn't wish for better treatment than was given me by Ruthie's daughter, Jackie. After 15 years, she and her family still give me the same admirable and appreciated treatment as a family member. They made it easy for me to adapt. This was a new home for me in a strange part of the country, where most of the people looked strange and talked funny. It wasn't like Texas, where everyone was a "neighbor" regardless of where you lived and were most welcome.

It seems to me that all the "natives" here have very little living space and have to fight for their jobs, surviving from one day to the next. There are too many people in a small space. They keep to themselves in little cliques and entertained family members only. They look with suspicion at unsolicited friendliness coming their way. They've had one boring job throughout their entire lifetime and that's all they can talk about over and over again.

Even the college children major in courses that are not marketable for good jobs. Lack of ambition seems to be the culprit.

Our fellow tenants in the Village apartments were all very nice, retired seniors and were all very friendly. We socialized with them and had some wonderful times together, but one by one, they went to nursing homes or were "called home".

Ruthie and I spent our honeymoon there and were very happy together. After that, it was back to reality. We went bowling at the Plymouth Lanes, where Ruthie was a member of the Monday Morning League, which consisted of thirteen teams. While she bowled, I rented shoes and a ball. There was always a lane open where I could practice alone.

After that first practice, Ruthie took me to a sporting goods store and we purchased a twelve-pound bowling ball, a pair of bowling shoes, and a bag to put it all in. We were both seventy-one years old!

For several weeks, I practiced while Ruthie bowled with her league. One day, one of the lady's didn't show up and there was no one to take her place, so they asked me to take her place that day. After that, they talked me into being a member for the rest of the winter season.

It was fun, especially when I dressed up as Santa Claus and bowled! We had a party and a few drinks while we were bowling. And, of course, there were a few ladies who had to have their picture taken on Santa's lap!

At the end of the bowling season, they chartered a bus to go to the casinos in Atlantic City, New Jersey. They sure knew how to travel! Two gallons of "screw drivers", ice, plastic cups, and several tomato pies!

When we arrived in Atlantic City two hours later, we were a happy group! There were thirty-eight women and me! Most of us won a little money on that trip.

I bowled as a regular member the next year. No one objected to my being there and they all treated me nicely. Once in a while, though, when one had an impossible split to make, I heard words spoken and put together that were unlike anything I'd heard before, even in my nine years in the Army Air Corps! I was amused and always pretended not to hear.

During the bowling season, we had to make a few trips to visit Ruthie's sister. Mae still lived in Hillsville and became incapacitated at times and we had to find caregivers for her. I did a lot of repairs on and around the house and mowed her lawn a few times. She had a riding mower, which I enjoyed. There must have been over half an acre of lawn.

Hillsville was mostly a retirement town for people who had worked in the coal mines in Virginia and southern West Virginia. They built their large dream homes of beautiful brick, nicely set back from the street with shrubbery.

Since it was on a plateau in the Blue Ridge Mountains, it was a fantastically beautiful place, always lush and green. It reminded me pictures you see on calendars.

They had two newspapers, The Carroll County News and The Roanoke Times. All the retirees would get up early to check the obituary column to see who had died. Then they would call each other because everyone knew everyone in Carroll County as well as the other surrounding counties. The phones rang a lot.

Everyone who was able, even those in wheelchairs, went to the viewing, held from seven to nine in the evening the day before the funeral. They would sit in the chapel and catch up on all the gossip.

The next morning, the phones would start ringing again and they would discuss how "Agnes" or "Blanche" or "Aunt Sue" looked so terrible and had been ailing lately. The old gossip was warmed over again!

Then the conversation would turn to "What are you going to wear to the service?" Everyone who could went to the funeral home for the service but not many went to the cemetery because it was quite a ways outside the city.

After the service were over, some went home, others went to the cemetery, then they all met at the home of the "dearly departed", where neighbors brought all kinds of prepared foods for the guests.

It was a custom handed down for generations and most people enjoyed the fellowship and comforted the bereaved family. It was a nice thing.

The following day, the phones would ring again and the conversation turned to what the deceased had left to each survivor. It seemed that everyone knew all about each other's financial status and property and material possessions.

Then the phones would become quiet until the next person died and the activity would start all over again.

Hillsville has about 2,000 people but after 6:00 P.M., the city looked deserted! There is no theater or any other sort of entertainment and Roanoke, the closest large city, was eighty miles away.

For an active person, it would be like living in a nursing home! But for those who had spent their lives in coalmining towns, it was Heaven. The higher elevation with dust-free air and wonderful climate worked wonders for those coalminers with "black lung" and other body ailments that were a result of years of hard labor. I could understand that!

CHAPTER TWENTY-TWO

Travels Overseas

A year after we were married, we decided we would like to take a fifteen-day trip to Europe. Don and Betty Ann suggested that we meet them at LaGuardia Airport in New York City and we could all fly to Geneva on a direct flight.

Don and Betty Ann had a chalet in the Alps near Geneva and invited us to be their guests there for a week. After that, we could tour Europe.

We found a travel agent who had made several trips to Switzerland, Austria, and Germany. She laid out a tour and made the hotel reservations for us from her Chestnut Hill office.

We made the flight over the Atlantic in May 1987. It was a smooth, uneventful trip and when we landed at Geneva Airport, Don rented a car, since they would be at their chalet for a month.

Their chalet was a dream home. The view from any window and from all parts of the yard was unbelievable, and highlighted by the clear, beautiful sunshine.

Ruthie and Betty Ann rode a cable car up to the town of Marian while Don and I took a cable train. It was breathtaking to ride the rim of cliffs hundreds of feet above the valley. We spent the day shopping and had lunch before we went to the Staubach Hotel, near the waterfalls.

We toured all the little towns around Geneva and learned a lot about the wonderful people we met there. My first impression was how clean everything was!

There were large white and tan cows with bells around their necks running free on the mountainsides. The only fences were around the vineyards. The cows were a great necessity to Switzerland's economy. And there was plenty of green grass for their grazing the instant the snows melted.

Don and Betty Ann were wonderful hosts. They took us all the places they could in that part of Switzerland. On our last day with them, they drove us to our hotel in Luzern, near a river fed by a large lake. When they left to go back to their chalet, we missed them immediately.

After we had checked in and found our room, we went back down to the hotel clerk to ask what we could do for the rest of the day. He advised us to go down to the dock and take the large boat across the lake to the little town of Regis. From there, he said, we could take a cogwheel train or a cable car to the top of the mountain. The view was spectacular, he told us.

We got on the boat and began our journey across the emerald green lake. From our seats on the top deck, we could see that the water was wonderfully clean and clear. When we reached the town, we decided to take the cogwheel train up the mountain. What a fantastic picture the snow-capped mountains surrounding us made!

On our way back, we got off the train about a third of the way down to take the cable car the rest of the way. We arrived at the station just in time to see a Swiss couple celebrating their wedding.

The bride was sitting in a straight-backed chair with a long pole fastened to each side. The groom carried the front part and the best man carried the rear. As they walked, they swayed the chair from side to side. They approached an area where two men were playing a cheerful tune on their 9-foot long Alps horns. It was a cheerful time for them and for us as well as we watched the performance.

After the wedding was over, Ruthie and I went inside the cable car building and boarded a car with a lot of other people. Our car went straight out to a tower high above the tall pine trees. Ruthie maneuvered us so that we could look out the front window. I looked down and there, two thousand feet below, the town of Regis looked like a postage stamp!

I used to suffer from acrophobia because, as a child, I had once ridden my three-wheeled wooden bike backwards off our front porch on the mountainside, down the many steps, coming to a stop and banging my head on some boards at the bottom. Serving in the Army Air Corps had helped me get over it a bit, but I still had some apprehension riding on that cable car out there in space! Once I calmed down, I really enjoyed the ride and the view.

One evening, we had made reservations at a dinner theater across the river. To reach the theater, we walked across a covered wooden bridge that spanned the river. Inside there were paintings on the ceiling and support beams that had been there for hundreds of years.

Our table at the dinner theater had a great view of the stage and we enjoyed a romantic candlelight dinner accompanied by beautiful music before the show. The show was fantastic and the cast received three standing ovation curtain calls when it was over.

On our way back to the hotel around 11:00 P.M., we noticed what must have been over a hundred and fifty bikes in the racks under the trees near our hotel. None of them were locked. We asked the hotel clerk about it when we went inside. His reply was, "Who would want to steal a bike when we all have at least one of our own?"

The next morning, we went back across the covered bridge with its ancient paintings to visit a museum with a large Picasso exhibit. It was quite extensive and I couldn't help but wonder what a person would have to drink to paint such terrible paintings! The only one I thought was nice was the first one he had painted.

Among the many places we visited was the Richard Wagner Museum. This museum was the composer's home from 1866 to 1972. There were many personal items, many original scores, pianos, etc., on display and there was a beautiful view of Lake Luzern from his home. When I saw his pianos, I thought of my first wife, who was not only a music teacher but a concert pianist as well. She always said that Wagner's music was a violent sort of music. Wagner had large hands and his music was written for his own use, so it was difficult for a woman to play.

We thoroughly enjoyed our stay in Luzern. It was a really fun city!

We left early the next morning headed by train to Munich, on schedule. It was a nice trip and the scenery was beautiful. Our hotel was located near the downtown section of the city so we had easy access to all the beautiful buildings, museums, and the lake in a park in the city. The Bavarian Kings, Ludwig I, II, and III had each left their mark by building palaces and various residences.

What really got my attention was how quickly the city had been rebuilt after WWII. I had seen pictures in our war rooms on our stateside bases of the city being leveled. By 1986, you couldn't see any sign of the bombing and devastation.

The nights in Munich were fantastic, all lit up with people enjoying walks and seeing all the places we missed during our daylight bus tours.

We took a train to the city of Dachau, which was nestled in a forest of beautiful pine trees. Once there, we transferred to a bus to visit the former concentration camp outside the city. There wasn't much left except for the administration building, which had been converted into a museum. There was still barbed wire, guard towers, and ditches around the perimeter.

The crematorium and gas chambers were there to visit. All that was left of the barracks was the concrete base that each had sat on with gravel inside to serve as a floor.

Most of the exhibits were a pictorial review of Hitler and how he got started in the prison life. There were also bins of shoes and clothing and personal effects that had been taken from the prisoners before they were sent to the gas chamber. There was a large picture of Hitler being congratulated by well-dressed men in Munich after his two and a half hour speech there.

It was a very thought-provoking place to visit and seemed very unreal.

As we left the crematorium, there were flowers blooming in the grass and a soft wind blowing through the pines. It was unbelievable that so many innocent people had died such horrible deaths here.

As we waited for our bus our by the highway in an open stand, we were having coffee when a young German man who drove a delivery truck by there every day began talking to us. He was very sad and confused that such a thing could happen in his country.

He felt that all the young people would have to pay for the sins of their fathers for years to come.

Our visit to Munich was short, but great, and soon it was time to board the train to Salzburg. I was really anxious to see this area because I had read that it was beautiful and I knew that the movie, "The Sound of Music" was filmed there. I had seen that movie at least three times!

Again, our hotel was in the center of the city but we took bus tours to see the sights outside the city. What a surprise! The Abbey where Maria began as a novice, the Von Trapp family home, the little summer house, and all the other places shown in the movie were miles apart! In the movie, all had seemed close together.

To top it all off, in the last scenes of the movie showed the Von Trapp family climbing the mountain to escape. In reality, they had ridden a cable car up the mountain, got off, and climbed only a few yards up the mountain! But it was still a fantastic story and in real life, the scenery was even more beautiful!

We had lunch on a open patio with a full view of a lake and an incredible view of the mountain in the background.

Outside the city of Berchtesgaden, high in the beautiful mountains, were Adolph Hitler's "Eagle's Nest Retreat" and Hermann Göering's magnificent home. Hitler's home had a seven by fourteen foot picture window built into the end of the huge living room. It was electrically controlled so it could be retracted so the fresh Bavarian mountain breezes could be enjoyed. The view was fantastic.

Allied bombing during WWII leveled both homes, although it was a difficult place to drop bombs because there were anti-aircraft guns hidden in the mountains around the area. After the bombing, there was only wreckage left. The first Minister President of the area, Wilhelm Hoegner, ordered the rubble completely covered with soil and grass to prevent any radical groups from causing new trouble.

Our tour bus took us to a parking lot at the foot of the mountain where we boarded a specially designed bus that would take us halfway up the mountain to the site of Hitler's Eagle's Nest Retreat. The front wheels of this special bus were recessed way back under the bus, as were the rear wheels.

The road was very narrow and I recall that as I sat in my seat behind the driver, I looked down and all I could see was open space to the valley floor. The Eagle's Nest and the road leading up to it are a masterpiece of construction and drew worldwide attention.

Martin Bormann, Hitler's right-hand man, wanted to give Hitler the Eagle's Nest as a 50th birthday gift but the costs involved required Hitler's approval, so it wasn't a surprise. Most people believed that Bormann built it to gain favor in Hitler's eyes and so he would become the second-in-command.

Construction was begun in the autumn of 1936, some 6,017 feet above sea level on the Kehlstein Mountain peak.

The bus took us to another parking area about halfway up the mountain, where we got out and went into a large tunnel, walking about 400 feet to a bright brass door behind which was a large, room-sized elevator. The guide explained that because Hitler was claustrophobic, the elevator had to be large and roomy.

The elevator took the entire busload of us up through the interior of the mountain to the large house Hitler called The Eagle's Nest. The view of miles of snowcapped mountains from the huge windows was indescribable and breathtakingly beautiful.

A large restaurant in the former living room served delicious food. I asked a lot of questions. The tourist folders placed the cost of the house at the equivalent of $90 million in 1981 dollars! It took more than 3,000 German and Italian workers, working night and day, more than thirteen months to finish the project.

Hitler used this retreat but not as much as people expected, because the altitude caused him to be short of breath. Eva Braun loved the Nest and spent a great deal of time there. In spite of the heavy bombing, somehow the Eagle's Nest was spared although the village below was leveled.

Then we were off to Salzburg, Austria and once again, our hotel was at the center of the city. We were within walking distance of a lot of the museums and palaces.

After you've seen a few palaces, they all seem to run together in your memory, yet each one had something different.

The Prince Archbishops represented the worldly and Catholic clerical power in the country. Their residences in the city of Salzburg were such that a number of impressive monuments, buildings, and jewels assured that they could live in splendor.

One such place was very beautiful with extensive grounds and gardens surrounding it. It was called the Hellbrunn Pleasure Castle and had been built by Bishop Marcos Sitticus in the early 17th century. He had a weird sense of humor and hired a fountain designer. Some of the fountains were tricky, giving your face a full wash! Others were hidden so that when you bent to read an inscription, you received a quick spray!

He had a large marble table built outside for his guests, with eight single stools also made of marble. Each seat had a small hole in the center that looked like a drain hole for rainwater, but hid a water jet. When the party became displeasing to the Bishop, he would signal the gardener from his royal chair, who would then turn on the water. This certainly got instant results!

The icy jet of water hit each guest in his "seat" with terrific force. Needless to say, they knew it was "time to turn out the lights and go home"!

The next morning we went by rail to Vienna. When we got off the train, Ruthie went outside to secure a cab to take us to our hotel while I gathered our luggage. She gave cab driver after driver the name of our hotel but they just shook their heads. Finally, one driver pointed across the street—and there was our hotel!

I had always wondered what the station looked like when my parents had stayed in Vienna three days and nights in 1906, waiting for their fake passports. And each day they went to a café near the station at noon and again in the evening waiting for the man to show up.

As usual, we asked the hotel clerk about scheduled bus tours and operas. He mentioned an opera at the Vienna State Opera House called "The Trial of Mary, Queen of Scots". We were able to secure two seats in a box that hung over part of the stage and we could see and hear everything very well. It was a wonderful performance. As Judy Canova, the lady comic who sang classical opera on the Grand Ole Opry radio show out of Nashville, Tennessee, once said, "Opera is when someone gets stabbed, shot, or poisoned and sings for two and a half hours before they die!"

I also enjoyed the intermission. We had wine in the huge hall, complete with seven large chandeliers and painting hung on the walls. It was also on the same floor as our box seats.

One day, Ruthie and I were walking in a plaza that covered a large area in the center of Vienna and we happened to go down a side street. We were both quite surprised to see a sign on a wall beside an arched door that read "Quessenburgh Palace".

Ruthie's mother was a Quessenberry. The name came from England and was derived from the Austrian name of Quessenburgh.

We tried to go inside but were turned away. It was a government building. Later, though, we found out that we should just go on through the archway and into the courtyard and enter through a back door. We did that and found it was a fantastic place with paintings, ornate furniture, statues, draperies, etc. We couldn't go inside any of the rooms because they were government offices.

It made Ruthie feel great to see where her family name had come from.

In the plaza, all kinds of people were selling their wares, musicians were performing, and the sightseeing was great.

We also visited a Spanish Riding Academy and watched the riders practice their intricate formations. The academy was a Spanish-style building with an earthen floor and a balcony with seats that ringed the riding area.

Then we visited the Theater of Fine Arts, Schonbrun Palace, which was a lavish 1,200-room summer palace that has changed little since Maria Theresa's era. There were only forty-five rooms open for public viewing, plus terraces, fountains, and a collection of Imperial coaches. It was a very elegant place.

St. Stephen's Cathedral was near our hotel and we could hear the huge brass bell called "the boomer".

One evening, we rode the world's largest Ferris wheel, which stopped each enclosed car at the top of the wheel, over two hundred feet above the city, where we had a wonderful view of the beautiful lights of Vienna. The ride was a prelude to an evening of a nice show at the Straus Park and of dining and wining! A professional dance couple danced a beautiful waltz and we toasted them with wine.

After the show was over, a tour bus took us to a winery to sample wines for an hour. Our group was composed of couples from different hotels and of all nationalities. At first, we couldn't communicate with each other very well, but as the wine flowed and the trio of musicians played, we became friendlier. And, of course, since there was an accordion player in the trio, we requested the Beer Barrel Polka. Everyone sang it in English a few times. It was a fun night.

As the bus was dropping passengers off at their various hotels, one German man said, "It is gut ve can talk more easy and frustee vat ve say after a little vine!" I agreed with him. We shook hands and I wished him well and we said goodnight.

The next day, we took the train to Frankfort, Germany and from there, a plane back to Philadelphia.

During the first several years, I didn't hear from my daughters but later, I received Christmas cards that did use the words "Dad" and "love". I sent the grandchildren, Wendy and Brooks, money for Christmas. After a few months, the ice melted some, but it was three years before they began to write in a more friendly tone.

Ruthie and I drove back to Austin, Texas a few years later to pick up some personal items and tools that I hadn't wanted to sell or give away. While we were there, we stayed at Don and Betty Ann's and they had a family get-together. Ellen and her family came as well as all of Don and Betty Ann's relatives. We had a good time.

My daughter, Barbara, didn't come even though she only lived about a mile away.

After fifteen years, the unfriendly situation between Ruthie and my daughters is the same. I always exchange letters and cards with my daughters and grandchildren.

Ruthie finally got to meet Ellen and her family but we weren't invited to visit them at their home in Bryan. Our route back to Philadelphia took us within twenty miles of their home. It was a sad and trying time for me, but being a survivor, I shut out the pain and, as usual in such a situation, thought of all the wonderful things that had happened in my life.

I made a mental note of plans to make my life even more pleasant and happier. I knew Ruthie, sitting there beside me, was disappointed. Her family and friends accepted me "warts and all!" as they say in Texas, but my own daughters would not accept her. All the Diviak's and their kin and all of my friends were really pleased and always greeted her warmly. One thing for sure—I was not going to dwell on the problem!

We arrived back in Philadelphia in a couple of days and it was "nice to be home again". Our nice apartment in the Chestnut Hill section of the city was a great place to live.

Ruthie belonged to a social club that was very active. They often took day trips by chartered bus to the casinos in Atlantic City, New Jersey, to dinner theaters, and so on.

Best of all, we had a group of nine or ten people who enjoyed each others companionship, so we often had dinners together, played penny poker and a Jewish game called Rummykub, which Ruthie and I still play. We leave the tiles on the table and when TV programs become boring or unbearable, we play. It does exercise the brain!

The main thing was we kept busy and never got into a rut or got bored. As someone aptly put it, "the difference between a rut and the grave is the dimensions"!

And all along, we were always invited to parties at Ruthie's daughter's home and often to parties given by their friends. We reciprocated, which gave us a lot to do. Never a dull moment and there is always something nice to think about when each day came to a close.

Our Christmas Eves and Christmas days were spent with Jackie and Hap and their children. We exchanged gifts and Hap prepared the feast we enjoyed so much. Then, a week later, we would treat them to a great dinner.

I was brought up with the idea that you invite people to your home that you love or want to get acquainted with or who were lonely and needed a friend. But as I grew older, I discovered that many people didn't share my view. They would dine with you yet never invite you to their home for so much as a cup of coffee.

I learned to still be nice to everyone, but if they didn't care to have us in their home, I never invited them again. I am a firm believer in "reciprocating" as we called it in the Army, and avoiding people who take advantage of others.

For the first ten years of our marriage, we took four trips to Europe and one to Hawaii and enjoyed every one of them. We had a wonderful time meeting new people and enjoying the countries we visited.

In the early spring of 1989, our favorite tour company sent us a booklet outlining a fifteen-day guided bus tour of England, Wales, and Scotland. Ireland wasn't included because of the strife there.

After all these years, they're still unable to get along with each other, while they still profess to be Christians! But that's nothing new. It has happened and continues to happen all over Europe and the Near East.

We left on May 25, 1989 from Philadelphia on a direct flight to Heathrow Airport in London. The flight took six hours and forty-eight minutes and was smooth and enjoyable.

The tour bus met us at the airport in London and took us to our hotel in Piccadilly Circus. This was a "rest day". We immediately check out the local sights not covered by the guided tour for those days we had free.

A professional guide came with us on the tour of Royal Albert Hall, Albert Memorial, Kensington Museum, Chelsea, Knight's Bridge with Harrods, The House of Parliament, Big Ben, the River Thanes, and a special tour of Westminster Abbey with all the regal setting for coronations since 1066. Then we saw the Changing of the Guard, which was colorful and very precise.

We had a free afternoon and made a tour of the Tower of London, where the crown jewels are displayed in heavy, bulletproof glass cases on the walls of the stairway leading deep underground. You weren't allowed to stop to look at them, but had to walk slowly past. There were guards there to make sure the line kept moving and no photographs were allowed. The display of crowns and jewelry was fantastic.

Then we saw Scotland Yard, and Buckingham Palace, but only from a distance due to security reasons.

Every morning at 8:00 A.M., our luggage was loaded aboard the bus and we proceeded on our trips. The first was the ornamental Garden at Henry the VIII Tudor Palace of

Hampton Court; then Salisbury Cathedral of early English Gothic design; the famous Stonehenge, which was unbelievable in its size, and the area it covers; and the Fleet Arms Museum.

We stopped overnight at Exeter and left the next morning for Cardiff, Wales. On the way, we got out of the bus to look at the small stone and concrete pier at Plymouth from whence Sir Henry Drake set sail for his adventures, and where the pilgrims boarded the Mayflower in 1620, bound for America. We visited the Dartmoor National Park, Glastonbury (King Arthur's Isle of Avalon), Wells Medieval Cathedral, and the Georgian City of Bath (Roman and pump room) built by the Romans when they occupied the area.

We went across a 3,240-foot long, seven-lane bridge into Cardiff, Wales. Every time I see the name "Cardiff" I think of the hoax that someone pulled off by creating a giant of a man, supposedly the largest man in the world, in concrete and passed it off to the public as a "petrified man"—the Cardiff Giant— in circuses all over America. They made a lot of money before it was finally proven to be a hoax!

Early the next morning, we drove to Chester, England. On the way, we saw the Brison Becon National Park—land of Salmon River, and Welsh Mountain Ponies. We visited a place called The Valleys, where iron and coal was mined, then the Island-Trinidad Wells, over Horseshoe Pass to the walled city of Chester.

We took a walking tour of Chester, past Roman ruins and the ruins of a cathedral. In the afternoon, we drove through the International Park where we boarded a boat that held all fifty of us easily. We cruised Lade Windermere and docked at another landing, miles away from where we had embarked. But our tour bus had made it and was waiting for us. Windermere is where Wadsworth lived.

Our tour took us to Glasgow, Scotland, Gretea Green, Maffat Tartan Tweet Center, and in the afternoon, to the Glasgow Festival Gardens, 100 acres of flowers and plants, a festival show, and street theater companies.

Our tour stopped at the “bonnie banks” of Loch Lomond in the highlands at Fort Williams, near the bottom of 4,406-foot high “Ben Nevis Hill”. Then it was on to Caledonian Canal, Urquhart Castle, and Loch Ness. Much to our amusement, we heard the story of “the monster of Loch Ness”. After seeing all the memorabilia for sale, we realized it must be a profitable business! After a short visit at Loch Ness, we were taken to our hotel in Inverness, the Highland capital. Later that day, we visited the local woolen mills.

I remember before going to bed at 11:00 that night, I pulled the drapes back and was surprised to see that it was still daylight outside! We were so far north that the nights were short. When we arose the next morning around seven, it was already getting light again.

We left Inverness and motored to Edinburgh Colladem Moors, Spey Valley to visit the famous malt whiskey distillery. It was a fully automated plant. We sampled the Glenlivet Whiskey being made there after we toured the plant. Everyone was amazed at the row after row of windowless concrete single-story storage buildings along the highway where the whiskey was aged in wooden barrels. There were acres of them in the fields.

We reached Balmoral Castle, Queen Elizabeth’s residence during the hunting season. We had lunch in a restaurant across the street from the castle. I had a venison burger and some ale, which was quite a lunch and tasted very good.

While we were there, we saw the Queen’s newest purchase of a couple of shaggy, longhaired cows that had come from the Highlands. The Queen was an active businesswoman among many other duties.

As we left on our bus, I looked out the window and saw a Land Rover driven by a lady approaching the bus on the right side. The bus driver announced that it was Queen Elizabeth herself! Behind her car was another Land Rover occupied by a couple of men in casual dress—her security detail.

The next morning, we toured Prince Street and the Royal Mile. We heard all about the "Body Snatchers", "Dr. Jekyll and Mr. Hyde", and "Greyfriars Bobby". Then we saw the Edinburgh Castle and the Holyrood Palace.

We had the afternoon off and that gave us a chance to visit the shops and see the people in town.

That night we had a Scottish nightclub party that included dinner, ale, wine, etc. I sampled the touted Scottish dish called "haggis", a famous national dish made from the heart, liver, and lungs of a sheep. These ingredients are chopped with suet (animal fat), onions, oatmeal, and seasonings, and then boiled in a bag made from a sheep's stomach. It's served on a platter, minus the stomach.

A few of the ladies soon quit sampling, but I'm a person who will eat anything, as long as it doesn't crawl off the plate. I thought it was quite good.

In my youth, my family raised and butchered hogs and made garlic sausage, liverwurst, and blood and rice sausage. In Texas, there is a famous "morning after" cure, when you've had too much tequila, called "ma-noodle" that you eat for breakfast. It's made of chopped tripe, onions, and broth cooked into a soup. After you eat it, you feel a lot better. Or at least you forget about your hangover!

At the nightclub party there were Highland dancers, bagpipes, and the Ceremony of the Haggis. The Master of Ceremonies was standing on a low stage. He announced that they needed male volunteers from the colonies to come onstage for a contest.

A young man from Australia, along with others from New Zealand and Canada went up to the stage. Another young man from the Northern USA went up, and Ruthie pushed me out of my chair to go and represent the Southern USA. We all clambered on the stage, to the cheers of the two busloads of people we had all come with.

Once onstage, we were each interviewed by the comical announcer. "Your name, please?" He asked me.

"Joe Diviak."

"That isn't Scottish?"

"No, sir. It's Slovakian."

"Where were you born?"

"In West Virginia."

"Where do you live now?"

"Originally Texas and now, Pennsylvania."

"Are you sure you're not Scottish?"

"Yes, sir!"

After he had interviewed each of us, he explained the contest. We were to pick up a scabbard with a sword in it, with a shoulder strap and belt, then a shield and put them all on, then draw the sword and assume a fighting stance. The announcer had a stopwatch and said that the one who could do that in the shortest time would win first prize.

I came in second and received a bottle of Glenlivet Scotch Whiskey and a scroll, authorizing me to form my own Scottish clan! It was a very lively evening! I love the Scot's sense of humor and friendliness. They sure know how to have fun.

We left early the next morning and visited the 50,000 acre estate and castle of the Duke and Duchess of Roxleurgne. Prince Andrew and Sara Ferguson were married in 1985. We traveled on to the house of Mary, Queen of

Scots, the Northumberland National Park, the Adrian Wall built by the Romans during their occupation of England, and York Minster Abbey.

We toured Yorkshire Manufacturing Center, Sheffield, Doncaster, Robin Hood's Sherwood Forest, and Ann Hathaway's House (she was the wife of William Shakespeare). Her home and birthplace were close to each other. Both places were very small but of course, they didn't have all the furniture we have in our homes today. And heating was easier when the rooms were small.

While we walked along the picket fence surrounding the Hathaway House, I noticed a small splinter of fence lying on the ground. I picked it up and we took a photograph of the house, then mailed both to a friend who has the Hathaway name. He really appreciated it.

We spent our last day in London, England after having covered 1,741 miles on our bus tour. We took the opportunity to visit a Medieval Market On the World, Churchill's grave on the grounds of the Palace of Blandon, Oxford, and Windsor Castle.

That night, we had a great feast at a dinner theater and saw the play, "Me and My Girl". It was funny and, of course, the actors were really great.

I learned a lot on this trip and found the people in all three countries to be very friendly and possessing a great sense of humor. A fellow tourist remarked, "They have to have a great sense of humor to have survived all the pain of wars, illnesses, and a severe caste system!"

However, I thought that perhaps it was in their genes and probably "survival of the species" because they were a naturally fun-loving people.

We rode the double-decker buses and found that it was easy to get around London either by bus or walking. We took a plane from Heathrow Airport and were back in Philadelphia eight hours later.

Once home, we resumed our bowling, parties, and attending the theater productions at Chestnut Hill's small theater. We made two more trips to Virginia to help Ruthie's sister Mae with some medical problems and to secure home care for her so that she could stay in the home she loved and not go to a nursing home.

As always, there were repairs and other work for me to do around Mae's home. I was still "Mr. Fix-it"! I enjoyed the visits with Ruthie's siblings and other relatives very much and they seemed to enjoy seeing us each time we were there.

We were fortunate to find a wonderful caregiver in the very beginning when Mae became unable to care for herself. She had many physical and medical problems, yet she still wanted to stay in her beloved home. As soon as we met Lois Alderman, we were quite taken by her. She is a beautiful, trim, neat, and hardworking lady. Lois had reared her own family and began helping others whenever she could as a caregiver. It was a blessing to us to find someone so gifted and capable. We still don't know how this lady manages to care for people who are usually quite a bit larger than she is. She gets them in and out of their beds to wheelchairs, and in and out of the car. She must have superhuman strength since she's so small herself!

Lois is neither a complainer nor a faultfinder. I have never heard her criticize anyone and she always has a sweet smile and a positive attitude.

She worked part time caring for Mae each week, and part time caring for Mae's brother, Guy Horton and his wife, Ella. After Mae and Guy died, Lois continues to care for Ella.

It made it a lot easier for us because we live over five hundred miles from Hillsville.

In June of 1992, we were traveling again. We took a plane to Copenhagen where we changed planes to Berlin. It was a beautiful, low-level flight and everything below looked so nice and green.

We became acquainted with our fellow travelers at the hotel. Since this was a guided tour, we were instructed to have our luggage outside our rooms by 8:00 A.M. and told that breakfast would begin at 7:00 A.M. We had found that this system really worked well in our previous trips to Europe.

Berlin was quite a surprise. All evidence of the bombing of WWII had been erased and a bright, sunny, beautiful country had emerged. Our bus tours both in and outside the city were very informative.

The first thing we noticed was a three-foot wide strip of the sidewalk that was reserved for bike riders. They were racing along in both directions at full speed! You could get hurt if you stepped over the line into the path of one of those bikes.

I was also amazed to see how slim and neat most of the people were in Berlin. The only "full sized" or overweight people were the tourists!

All of the palaces, churches, and other marble buildings had been so neatly repaired it looked as if they had escaped the bombs. Other buildings were new and of modern construction.

Our tour bus took us to the Olympic Stadium, where Jessie Owen had beaten Hitler's best and won his Olympic medals, much to Hitler's anger and didn't say much for Hitler's idea of a "super race" either!

We went through the Brandenburg Gate and Checkpoint Charlie, which had been left, along with part of the Berlin Wall—the Iron Curtain—for future generations to see.

Our guide told a story of a very wealthy man and his wife who drove up to the gate in their plush, expensive car. When the guards asked the man if he had any contraband, his smart, egotistical reply was, "No, but maybe a bomb!" The man and his wife were immediately asked to get out of the car and the car was pulled over to the side. The guards proceeded to remove everything from the car; the luggage was thoroughly searched and everything left in a pile; the tires were deflated and the seats all removed. Finally, the border guard captain said, "We found no bomb. You may leave." The man was indignant and asked, "Who will put my car back together?!" The guard merely answered, "That's your problem!"

The moral is to never try to be a wise guy when you visit foreign countries. They don't have the same laws we have and can put you away for a long time!

When we entered East Germany, we were shocked to see the difference in the looks of the buildings. All the buildings in East Berlin were dull looking and in need of repairs and paint and cleaning, while in West Berlin it was like a whole different world—everything was bright and neat.

I was also quite surprised to see the area where Hitler and Eva Braun's bunker had been. There was no bunker there now, just a flat, wide, grass-covered strip of land, much like the farmland in our Midwest. From the pictures I had seen in our Army Air Corps war rooms after WWII, I had envisioned a hill or knoll at least.

One afternoon, after we had returned from a tour of the city, Ruthie and I and a couple from New Mexico decided to find a bowling center. There was one listed not far from our hotel and we found it easily. Once there, we rented shoes and a lane and started bowling. It was similar to our bowling centers in America, but the instructions on the consoles were in German. We had no problem though. There were only twenty lanes and the people bowling near us spoke English and were quite surprised to learn that we bowled in large leagues in America. They were nice, friendly people. And, yes, you guessed it—Ruthie beat all three of us with a 180 game!

Our tour covered the zoo with its zoological gardens touted as the world's largest, the Egyptian Museum which had objects from every dynasty in Egypt through the ages, and the most famous street in East Berlin, Unter den Linden (under the lime trees), lined with mansions and beautiful trees. The one and only wrecked church nearby was left for future generations to see.

Then we were off to Dresden and Budapest, traveling along the Iron Curtain Wall, which had all sorts of paintings on it and a reminder to future generations of what can happen in wars. Most of the artwork was quite good, including one showing the former East German President kissing the Russian President rather passionately on the lips!

We stopped in Dresden for lunch. Our guide told us that the people of Dresden were still bitter because of the late bombing that took place the day the war ended. I understood that the place made a lot of optical equipment for the military and was a viable target. It sure made believers out of the residents—they had felt quite safe during the war.

We visited several museums but the main one was Porcelain Collection that dated back several hundred years. We went to the Old Masters Art Gallery Museum, which was near there, then back to our bus and on to Budapest.

As we rode down the highway, it was hard to believe we were in Europe. The highway, guardrails, and road markings were the same as in America. Even the beautiful forests and plowed fields and the countryside looked the same, except that the barns and houses seemed to be attached for convenience in the winter.

Budapest, Hungary, is known as the "fun capital of Central Europe". It was one continuous party once we had seen all the museums, churches, and other sights.

The first evening, we went to a dinner theater and were delighted to hear the Gypsy music as well as the Hungarian chundiashes. My favorite was a huge zither-type string instrument that resembled a piano stringboard, laid horizontally on folding legs. It was played with two short sticks with padded bird-like mallets on each end. It must have been the forerunner of the piano. I had heard and seen them before, at Hungarian parties, along with violins and accordions. It brought back memories of my youth.

We took a trip to the little village of Holloko, which is a showplace of Hungarian lifestyle seventy or eighty years ago. The women wore dust caps, multiple layers of skirts, colorful blouses with a lot of embroidery, and boots.

The men wore laced embroidered vests over shirts with puffy sleeves and shiny boots.

The houses were small but comfortable looking. There were no modern conveniences. The people sang songs and performed folk dances. It was a real treat for me since I had seen some of the dress and dances from our Hungarian neighbors in my childhood in West Virginia.

There was a tiny church that would seat about twenty people used by the villagers and visitors were welcome. The so-called "Chain Links Bridge" that spanned the Danube River fascinated me. Instead of huge cables supporting the large bridge, a chain similar in design to a bicycle chain was used. The links were about six feet long and the steel plates were at least two inches thick and eight inches high. It was quite impressive with its masonry or concrete towers through which the chains passed. It was the only chain-supported bridge we ever saw in Europe.

After three days, we were on our way to visit Prague, Czechoslovakia, a city I had always wanted to see. I was thrilled to be there! My parents had talked about it often, but in my mind, I always saw it in black and white. There were no pictures of European cities when I was a child.

Our bus picked up Professor John Lovas from St. Charles University in Prague. He was to be our guide in Prague and through the rest of that country. I was surprised at the tremendous size of the city—approximately 1,300,000 people lived there and it covered a huge area.

The St. Charles University was the first built in that area, by King Charles IV. He was credited with saving Martin Luther from being tried as a heretic. Then he tried to keep the emissaries from Rome from selling indulgence letters of forgiveness to the Roman Catholics in Central Europe. Each letter permitted forgiveness before the sins were even committed at the time the member submitted it to the priest at the confessional, if the committed sin was listed in the letter. The money was needed to build St. Peter's Basilica in Rome.

Ruthie and I, along with the rest of our tour group, took a cruise on a river boat on the Ultava River which winds its way through Prague. What a wonderful view of the city and the beautiful homes along the river's shore.

We walked across the Charles Bridge, built by order of King Charles IV in 1350. There are thirty statues of religious figures on the walls of the bridge. It was an awesome sight to walk by them as you crossed the bridge.

An astronomical clock on the south side of a tower in the city hall showed not only the time, but days, months, holidays, the longest and shortest days, etc. The clock was made in 1410 and rebuilt in 1490 by one Master Hanus of Charles University. Legend has it that when he finished, the Municipal Council had him blinded to prevent his building another clock like it in some other town. It is said that after that, Hanus supposedly climbed the stairs in the bridge tower, stopped the clock, and then jumped to his death.

The clock remained silent thereafter until Jan Taborsky restored the mechanism to working order in 1552-1572. It is an unbelievable clock and incredible to see all the things it displays.

We took the tour bus for a day in Bratislava, which I was looking forward to visiting. It was the largest city near where my parents came from, only fourteen miles distant from their hometowns of Borsky Mikulas and Bursky Peter, both of which I also wanted to visit if possible.

As we drove along the highway, I noticed that all the small towns along the way had been built around a church. The houses had fenced yards and gardens in the back and along the sides and a few fruit trees. Everything was very well kept. There was no livestock or barns visible in the little towns. Then I remembered my father telling me that all the farms with their barns and cattle were out in the countryside. The towns and village clusters were for mutual protection, a holdover from all the wars they had suffered through the years.

I asked the Professor, who was still with us as our guide, how I could get to the two towns so I could see where my parents had come from. He left, then came back and told me that he wasn't sure the towns were even still there because of the two world wars and Communist rule. All the small farmer owners and farm workers had been moved into the towns to work in the mills or on the co-op farms.

He told us that it would be a bit risky to go looking for my relatives at that particular time, when the Czechs and Slovaks were peacefully separating from each other. There was still a lot of Communists embedded in the system who could cause trouble to many people.

A cab driver told us the same thing. No one would take us there, so I had to finally give up that idea.

It was nice to enjoy once again the food I had grown up on and the wine was as good as my father had said it was and so was the Pilsner Beer. The chicken-fried veal, which was dipped in a batter, was the best, along with potatoes and gravy, and so were the pastries, like kalachies, with the coffee. I knew how to make stuffed cabbage (pigs in a blanket) and such, so I opted instead for food I hadn't had since I left home in West Virginia.

After a few hours, we were on the road again, on our way to Vienna, where we spent the night. The next day, we boarded a plane bound for Copenhagen, Denmark, with others of the tour group who had, like us, paid extra for a three-day tour of Copenhagen. The others continued on to New York.

In Copenhagen, I was a little disappointed in the Tivoli Gardens, touted as the greatest in the world…it was just a bunch of animals in cages and rides for children, some fountains and gardens, and a lot of commercial stands. It was about a thirty-minute tour.

We learned to ride the buses to the places we wanted to visit or we would walk from our hotel. We had special three-day passes to ride the buses and train.

We took a boat ride to the Mermaid Statue across the harbor, but on the way, a thunderstorm came up suddenly and we had to rush to the shore near the Statue, which is on a rock. After the storm passed, we took a cab back to the hotel.

We visited Christian Anderson's home and museum near our hotel. The Museum fronted the harbor. His books and memorabilia were on display.

We found a bowling center and enjoyed a few games. It was a long walk to the place, but a nice walk, with flowers and trees along the way.

Ruthie and I, along with two other couples from our tour, were dinner guests of a couple who lived in Copenhagen. They had a nice home and we enjoyed the four-course dinner. He was a newspaper reported and she was a kindergarten teacher.

I was amazed to find that even those who owned automobiles chose to ride their bikes to work. The land was flat, with no hills at all, and gasoline was over three dollars a gallon!

We boarded a train with our special passes to visit Helsingor, a city in northeastern Denmark. This is where Shakespeare chose to cast the setting of "Hamlet". The Castle Kronberg is every bit as impressive today, as it was when it was chosen for "Hamlet".

Our two friends from New Mexico were with us and we all took a large ferry across a large bay to Halsingborg, Sweden, on the Orsund Sea. We saw the Karmon, an 11th century tower fortress. Nearby was Krapperip, the castle residence of the Gyllentlena of Shakespeare's "Hamlet". We had lunch there and stopped in a number of small shops along the Channel.

I was surprised to hear so much English spoken everywhere we went in Denmark and Sweden. On our train trip to the ferry, I asked a boy of about twelve if he could speak English. "Yes, sir," He replied. "Very fluently. We are required to speak English in the fifth grade in school. I was amazed! Not once did I hear, "you know what I'm sayin,'" or "you know, you know" or "you know what I mean", which is the language of most of our athletes interviewed on TV before or after games.

We had a late lunch before boarding the ferry back to Helsingor to catch the train back to Copenhagen.

Once again I noticed that all the Danes seemed to be trim people. And again, the tourists were the only overweight people we saw there! I had never seen twin or three-seated baby buggies before, but they were being pushed along the streets everywhere we went there…well, they do have long cold winters there!

One of the favorite sports in Denmark is backpacking and hiking and camping. The train was full of people loaded down with gear all going someplace to enjoy their favorite sport.

All the islands there looked so nice and green and I kept wondering how it would be to fish in one of the coves.

Ruthie and I and our friends visited Christianborg, which is the seat of the Danish Parliament, foreign offices, royal reception rooms, and the Museum of Danish History. After spending three days, we boarded a flight to New York and Philadelphia.

Ruthie and I lived in the Chestnut Hill Apartments for three years. The apartment was near a nice shopping center, a bank, post office, and a deli. But after a while, our friends began to pass away or move to retirement homes. The businesses in the area began to close, and there were quite a few muggings by people not from the area.

We began to watch for houses for sale, but most of them were old and in terrible condition, yet very expensive. Most of them didn't have a garage, or even a driveway. The only parking available was on the street. There were too many homes jammed together on both sides of the streets, and most of the people already there had two cars per family.

We finally found a townhouse in a nice, clean neighborhood. After we purchased it, we did a lot of painting and remodeling before we moved in.

What surprised me the most was the fact that, at age seventy-four each, after giving them our down payment, we got a thirty-year mortgage with no problem!

After moving in, we had to have lights put in closets, parts of the basement, and on the back of the house. One of the previous owners had a 16- x 20-foot concrete slab poured in the back of the townhouse, but it had no roof over it. We added a sixteen-foot square gabled roof with skylights over the concrete slab, then decided we needed a smoother floor.

We covered the floor with a lovely coral ceramic tile and it became a fantastic room out there! Later we added a six-foot high lattice fence alongside of the chain link fence that was already there, for added privacy. Then we added clear plastic storm screens inside to keep the rain and snow off the tiled patio.

The open end of the patio faced a grassy bank that was difficult to mow so we hired a landscaper to remove all the sod on the 14- by 16-foot bank. They replaced the sod with stepping stones, boulders, and river rock and planted shrubs, plants, and flowers, all according to the plans Ruthie had drawn up for them.

Since then we've added roses and crawling roses, impatiens, among other flowers. It's a fantastic place for cookouts and cocktail parties. We wanted that place to be able to get outside and read, drink coffee, or just relax and enjoy the beautiful plants. All our friends love it!

I decided I wanted a bathroom in the basement since there was only one bath and it was on the second floor. The garage is in the basement, which is at street level in the front of the house. Actually, the house is three stories at street level in the front, but only two in the back because it's dug into a hill.

In back of the garage is an iron sewer pipe that had a crack in the back and once in a while, water would seep from the crack. We called a plumber to repair it. He told us he would have to jackhammer the concrete in order to change the pipe. We asked him to put in drains for a shower, commode, and lavatory while he was repairing the sewer pipe. He did that and added copper pipe for all three units and he did a nice job.

I finished the walls in the bathroom after the fixtures were installed. Installing the bathroom cut our garage to only fourteen feet long, so I drew up some plans to extend the garage out to the front of the front porch. A contractor moved the garage door and the basement door. We gained a full-size garage and a second bathroom, much to our joy.

Years later, we had an aluminum awning installed over the front porch to keep the rain and sun off the front windows.

When we had first moved here, we wanted to carpet the basement, so we went to a carpet company and bought enough to finish the entire basement. Now we have a treadmill, and exercise cycle, and I built an entertainment center with a TV and stereo, so that I can exercise and watch TV. We also installed central heat and air conditioning, which was easy since all the ducts were already in place for the old-style gas furnace that had been used for years.

Now we have a comfortable home with little upkeep needed. I added an electric garage door opener, which is one of the best things we added to the house, especially when it's raining or snowing.

Once again, we began to get "itchy feet" so we took one of the tours to Italy that was offered at a very fair price. After a lot of preparation, we had everything taken care of and took a limo to Kennedy Airport in New York City and boarded our Alitalia 747 for the overnight flight to Rome.

The tour director met us at the Rome airport and the bus delivered us to our hotel to relax the rest of that day. A large group of us had dinner together at the hotel that night.

The next morning was spent touring Rome in the bus. The afternoon was free and we went sightseeing on our own, to the Colosseum, St. Peter's Square, the Vatican, the Roman Forum, and the Catacombs.

That night, we had cocktails on the Biancaccio Palace terrace, then dinner and dancing in the ballroom with a live orchestra. I couldn't help but feel great about the fact that here I was, a poor coalminer's kid who grew up in the Appalachian Mountains, dining and dancing with my wife in such an elegant place! There were a lot of huge crystal chandeliers, paintings on the ceiling, hugs paintings hung on the walls, and incredible furnishings. The old song, "You've Come a Long Way, Baby", kept running through my mind.

One of the Palace's heirs lived on the second floor. He was 94 years old. He allowed a few tour groups to come through to help him with the expenses. Our tour company had exclusive rights for its groups.

The fourth day, we were again on our own, and went to various museums and other sights by walking in different directions from our hotel. We had lunch and dinner nearby.

The next morning, our tour bus loaded with our luggage, left at 8:00 a.m. for a beautiful drive south to Sorrento via Naples, the birthplace of Caruso, Sophia Loren, Bell Canto, and "O Solo Mio".

The highways were much like ours, similarly paved and with guardrails. The scenery was beautiful, with all the vineyards.

Naples was a busy city full of traffic, with ticket sellers, cabbies, shoeshine boys, hotel hawkers, and wristwatch and pocket pen peddlers. Naples is a crescent city, forming a kind of an inward arch along the shores of the Mediterranean Sea. It is a beautiful, white city as far as the eye can see along the shore. We motored on to Pompeii to visit the ruined city which had been dug out of the twenty-five feet of ask that covered it when Vesuvius erupted in 79 A.D. Only thirty percent of the city has been excavated but it was quite a disaster. Some 30,000 people had been in the city when the volcano erupted.

I asked our tour guide what they had done with all the volcanic ash when they excavated the ruined city. He replied that the glass manufacturers had used it to make glass and some had been used to make mortar for buildings. My mind raced back to 1948, when a building contractor in our Businessmen's Bible Class back in Bryan, Texas, had made a trip to Rome.

He said that the most amazing thing to him was how a forty-step set of stairs without any back support could stand for 1,000 years in those ruins without cracking! Later I found out that the builders had added one shovel-full of volcanic ash to each shovel-full of cement, sand, and gravel to make the concrete. The ash is so fine, much finer than sand, that it keeps the water out of the concrete once it is set.

There were bodies covered with the ash, which were kept intact and later duplicated by using plaster of Paris over the bones to form full-size bodies. They were in different positions when they died of fumes or suffocation.

The Romans built the city 300 years earlier and had buried sewer ducts under the gutters in the streets. It was the first city with a sewer line to a large holding tank the size of a lake, away from the city. All the streets were of cobblestone and nine-foot wide sidewalks fronted the homes and businesses.

The first thing restored was the public bath and toilets. It was quite large with long rows of seats next to each other. One painting framed on the wall had survived the volcano. It is still there. Another building that had been restored was a small two-story brothel, with its six marble slab beds. It had almost escaped intact. The brothel was situated conveniently close to the business district, on a street leading to the seaport.

We drove past the volcano itself. I was quite surprised to see the beautiful homes and even a monastery built on the slopes of Mt. Vesuvius, along with a number of vineyards. I couldn't believe my eyes because, according to the people we talked to there, it could all happen again!

On our way back to Sorrento, one of the most flowered cities in Italy, we went on the high cliff road, the Almalfi Drive, along the bay. It was a bit frightening to look out the bus window and see nothing but water far below, with only a low rock wall between you and the sea!

Our hotel in Sorrento was on the high cliffs overlooking the Mediterranean Sea a couple hundred feet below a low concrete and rock wall. I looked out over the wall and saw a chain link fence type of net suspended out of sight. If anyone fell over the side, the net would catch them.

Across a cove from the hotel stood a huge mansion with an elevator inside a solid rock cliff to the boat dock and shed, where a large white boat stood at anchor. The sea was such a bright blue and so beautiful! And there were semitropical plants and shrubs and flowers everywhere you looked in Sorrento.

On our second morning there, the bus took us to the harbor, where we boarded a large boat to the Isle of Capri. I was excited to be seeing this place I had heard songs about on the record player in the forties and fifties. I wasn't disappointed. It was as romantic looking as the songs had led me to believe!

As soon as we docked, about twenty of us per boat cruised around the back of the island to a large cave in the cliff with the opening to the cave arched very low in the water. It was the "Blue Grotto", where it was rumored the nobility came to swim and bathe.

We left the larger boats and were loaded three of us to a small rowboat with an oarsman. He waited for the right time, when the swell left the entrance, and propelled our boat inside as we lay in the bottom of the boat. Inside, the Grotto was a huge domed room with a fantastic blue hue. The clear blue water was reflected on the ceiling. It's easiest to get into the Grotto when the tide is out and various tour people circled the area in small rowboats, waiting to take tourists inside. To get back out of the Grotto, the oarsman waited until the trough of the wave came in, then he propelled us through quickly.

We returned to the pier where our large boat had docked and boarded a bus to tour Capri. We had a wonderful lunch in a nice restaurant there, and then we walked a long way around the top of the island to see the beautiful foliage and the sea. We took a cable train back to the dock, then the boat to Sorrento, and the bus back to the hotel.

That night, we enjoyed a great dinner at the finest restaurant in Sorrento. The wines, as usual, were different than other places we had visited. Each area had vineyards and the competition was keen, each bragging that their wine was the best!

The following morning, we boarded our bus and headed to Florence. The drive was a long one but very beautiful and quite enjoyable.

We had a delicious dinner at our hotel that night and the next morning, (October 9, 1993), we rode the tour bus to do some sightseeing in Florence. It was here at the cathedral, in 1501, that Michelangelo found an enormous block of marble abandoned by another sculptor. He had "ruined" the block while trying to sculpt a statue of a giant.

Michelangelo signed a contract to execute to perfection this figure within two years. He turned the slab into the now famous Statue of David, much to the astonishment of everyone who saw it.

When we saw the statue, I was amazed at the facial expression and the expression in the eyes. The entire body looked so real, even down to the veins in the man's arm. We were told that Michelangelo had gone for years to the medical school at night to examine cadavers to see how the muscles were controlled and how they tied to each other. He made many drawings and notes so he could "put life into his statues."

I had read of a flood in recent years on the Arno River, which flows through the city of Florence. Apparently, museums and art storage facilities were among the buildings that were flooded. Many people from all over the world, including America, had come to help dry and restore the artwork and other valuable artifacts damaged by the floodwaters, so that they were saved for future generations. Florence is a city filled with paintings, statues, and museums filled with the famous works of many besides Michelangelo.

We walked along the banks of the Arno River. It wasn't that wide, but was now contained by concrete walls on both sides, not just marsh banks as it was when Florence was a city on only one side of the river.

I thought of a story I had heard forty or forty-five years ago: In Florence's early days, it was a busy little port on one side of the river. Soon, a few merchants decided to establish their businesses on the other side of the river. A ferry was used to transport them back and forth across the river. The King was at a loss as to how he could finance the construction of a wagon and coach bridge across the river, since his people were already overburdened with taxes.

A prince owned a palace with huge grounds surrounded by stone walls in the city near the river and the entry and exit gates to his palace were both well guarded. This prince had an interest and perhaps investments in the commerce on the far side of the river.

One day, a large number of posters appeared all over town. "NEXT FRIDAY, AT 10:00 A.M., FREE CLOTHING WILL BE GIVEN TO THE BEGGARS THROUGH THE GENEROSITY OF THE PRINCE."

Sure enough, when the time came, hundreds of beggars were permitted to come inside the palace walls. They were made to enter in a single column and line up, row after row, on the lawn of the palace and told to stand at attention. In front of each one was a pile of clothing. The guards gave the signal for them to undress and don their new clothes, leaving the old ones on the ground at their feet.

Then a band struck up a lively marching song and each column of beggars marched past the prince's reviewing stand. He smiled and waved at them as they marched past and out the exit gate. Once they were all out, the guards locked the gates and began to search the discarded clothing. Sure enough, they found enough valuables and money, not only to pay for the beggar's new clothes, but to build the bridge across the Arno River as well!

The bridge was named "Ponte Contra-Denti", which means "Beggar's Bridge". For years after that, the beggars and their descendants would look at that bridge and curse it!

I looked for the bridge, but all the bridges were modern steel. None were of rock and mortar with arches, as they had been built in earlier centuries.

We toured the Grotto Belfrey and the Baptistry. There was so much to see and to learn here, about Michelangelo, Dante, Da Vinci, and others.

We stopped at Pisa and visited the famous tower, leaning as if it would fall over. And, like most tourists, we found the correct angle to take a picture of me holding it up. The tower was fenced off because workers had placed wide steel bands around it at each floor level and attached cables to prevent it from shattering while they prepared to begin using hydraulic pressure to stop the leaning and gradually make the tower partially right itself.

It seemed to me that it would make more sense to tunnel under the tower from the opposite side from which it was leaning and slowly remove a few inches of soil and rock from under the high side, and let it right itself in a few years. There is no quick fix.

From there our tour bus took us to a dock, where we left the bus and boarded a large ferry on foot. We watched from the second level deck of the ferry as our tour bus was driven onto the lower level for the trip to a large island across the Canal Gladecca from Venice.

After we docked, we again boarded the bus and went to our hotel. The hotel was magnificent and very well located. And the grounds were beautiful.

After breakfast, our tour bus again took us to the docks, where we boarded boats for a short trip across the Channel to Venice. Once we landed, we walked a short distance on the sidewalks running along the edge of the Channel, and then across the "Bridge of Sighs", built in 1610. Prisoners were walked across the bridge from the prison to the front of the palace. The king would sit on a throne on the second floor balcony and watch them cross, and would judge which prisoners would live and which would die.

If the king gave a "thumbs down" sign, the prisoner was taken a short distance, perhaps twenty steps, toward the Channel, and beheaded.

We visited museums, the glass works, and took a gondola ride up and down the canal. An accordion player rode in one of the three gondolas and played as we all sang (?) the songs. It was fun but you could see that the rising water had begun to take a toll on the bottoms of the buildings. The water was a murky green and had a lot of trash floating in it.

It would have been a more romantic ride if it had been in the evening with all the lanterns rather than in broad daylight.

We spent two days visiting all the sights and marveled at how they had built a city the water that averages eight feet deep.

St. Mark's Cathedral was a case in point. The church and square was built on 1.6 million tree trunks that had been driven, side-by-side in one big square, by pile drivers until they hit bedrock. This was then capped by concrete after the tops of the posts had been cut even, about four feet above the water. As time passed, the trees became petrified due to the high mineral content of the water. There were dozens of islands in the area, made in the same way, all with large buildings on them.

The next morning, we returned by ferry and boarded our bus for a ride on the mainland of Italy. Our first stop, seventy miles from Venice, was in Verona, the home of Shakespeare's "Romeo and Juliet".

We walked through a street-level passageway into a courtyard that was partially paved with cobblestones. It was a beautiful place, hidden from the street, and boxed in by buildings. There was a balcony on the second floor of one building, overhanging the first floor windows. It was a small balcony with a railing around it, near the left corner of the house. There was a creeping vine similar to the ivy we see on our college campus buildings in America, and some blooming shrubs to the right of that.

Looking up at the balcony brought instant replays in my mind of Shakespeare's plays I had seen in America. But more than that was the memory of a nice, elderly Literature and English teacher I'd had in tenth and eleventh grades. Mrs. Eubank had tried her best to instill some culture into a bunch of us "hillbilly" kids in the Big Creek High School, whose only knowledge was of mountain folklore, such as the feud between the Hatfield's and McCoy's, and coalmine explosions and such.

I could picture her walking back and forth in front of the class in her wool plaid pleated skirts, turtleneck sweaters, and black Cuban-heeled shoes, her hair done up in a bun on the back of her head. She would dramatically read and quote all the plays and phrases she wanted us to remember. Of course, each one of us boys had our own idea of what Juliet might have looked like and it certainly wouldn't have been what Mrs. Eubank would have envisioned!

One of my classmates, Everett Moore, was a born comic and was always making up poems that were quite good. One day he recited a poem to us boys that he said was a quote from Shakespeare when he saw his first cowboy: "What manner of men are these that amble along with testicles dangling in parentheses?" We boys loved it!

A few days later, Mrs. Eubank was writing some material on the chalkboard and when she reached a part that required parentheses, she turned around and said, "Everett, these are also called 'bow legs' by the uneducated!" The whole class burst out laughing and poor Everett's face was red as a beet! Everett calmed down for a while after that!

From personal experience, I could understand Romeo and Juliet's problems. Her parents felt that she could do better both socially and financially and marry someone from a prominent family, and not someone like Romeo. And, of course, that would also upgrade their status in Verona. The families of Romeo and Juliet were also at odds with each other, which didn't help either. Romeo and Juliet ended it all with their deaths.

In my case, LaRue and I married without their blessings, which was a far better way out! I still have a warm feeling about the fortitude we had to go on and be happy together rather than be apart. We shared a wonderful life for forty-five years before LaRue died.

We were surprised to see a McDonald's restaurant of a design that fit nicely into the motif of expensive-looking buildings that were all around it. It was nice to have burgers and coffee there!

Then it was on to Milan, where we visited the Castello Sforzesco, a fifteenth century castle full of priceless artwork, tombs, and sculpture. Michelangelo's Pietά Rondanini was fantastic!

We visited the Duomo (cathedral), the largest in Italy and fourth largest in the world. It has over 1,000 spires and 3,000 statues both inside and outside the cathedral. It was planned entirely with drafting charts material list 500 years before it was completed. It seats 25,000 and can accommodate up to 30,000 with standing room in the main part of the church.

Being a bit of a "building nut", I tried to see where each successive generation of workmen and architects had added on to the previous building. There were no overlaps, cracks, or trims to hide the new additions. To me, it seemed fantastic to work on one building for five hundred years and do everything so precisely. It helped that all the building materials came from the same source, too.

We also visited the Teatro alla Scala Opera House, commonly known as La Scala, which first opened in 1778. It was heavily damaged by bombs during WWII, but reopened in 1946, under the baton of the famed Arturo Toscanini.

Finally our tour bus took us to Lugano, Switzerland, where we visited the Italian section to shop and walk along the shores of Lake Lugano. It was a beautiful ride along the lake, with the expensive looking homes and hotels along the shore in the Italian Alps. The next day, the fourteenth day of our trip, we left Milan and flew to New York, then on to Philadelphia.

Back at home, we immediately fell back into our routine of bowling with two leagues each week, and exchanging dinners with Ruthie's children on the weekends, along with attending a few plays.

I also had the job of updating the Diviak family tree and the minutes of the Diviak Family Reunion, which was held every two years, on the third weekend in July, in Abingdon, Virginia. I had to make forty copies and send them, along with pictures, to the lead members of each family. They would then make copies for their own children.

We had a few cocktail parties on our patio and cookouts for our friends and Ruthie's family.

Ruthie's daughter Jackie's family 1999
Left to right: Jackie, Harry, Ian, and Laura Gwyn

In October 1995, we decided to take a trip to Hawaii. We flew to Atlanta, Georgia and boarded a jet for the eight-hour direct flight to Honolulu, Oahu. When we arrived in Honolulu, we were greeted with leis at the Sheraton Surf-rider Hotel, at a Welcome Bar cocktail party. It was a fantastic party, complete with Hawaiian music. Later, we took the leis back to our hotel room and placed them in the refrigerator.

Later that evening, we had dinner and wine with our fellow tour group members.

The next morning, after breakfast, we had a "familiarization talk" about what we would be seeing that day. The rest of the day was free.

I went for a swim in the ocean in front of the hotel on Waikiki Beach. The clear water was cool and the waves were just right for swimming. Later that night, we had a great buffet dinner followed by a musical group known as "The Society of Seven". They also did impersonations and comical music. It was a great evening!

We had a morning of leisure, then a tour that included the Polynesian Cultural Center with its South Pacific villages and arts and crafts. Then we enjoyed a tropical buffet and an evening show.

On Sunday morning after breakfast, all the WWII vets put on our leis and took the tour bus to the site of the Arizona in Pearl Harbor. We assembled in a theater where a movie about the Japanese attack on Pearl Harbor was shown. It gave a fairly accurate account of the bombing. Then we took the ferry from the pier near the theater to the Arizona. A platform of concrete and marble straddled the sunken battleship and you could see the rusted line of the deck and hull. All the superstructures had been removed. Inside a room in the Memorial was a large wall inscribed with the names of the men entombed in the hull below.

We all stood there for a few minutes, as if in prayer, then went back outside and looked down at the sunken ship. The man next to me removed his lei and dropped it in the water. I did the same, as did all the others, one by one. I noticed a couple of Japanese ladies dropping their leis, too, and wondered if perhaps they had also lost someone in the battle of the South Pacific, in the American Armed Forces.

I had a chance to look at the other seven markers in the harbor where other ships had sunk that day. Then I began to wonder why and Commander would put all the ships in such a small harbor with only one entrance. If just one ship had been sunk in the entrance to the harbor, it would have bottled up all the rest. It must have been like shooting fish in a barrel for the Japanese dive-bombers! One story had circulated in the newspapers, saying that President Roosevelt had wanted them there so that the Japanese government couldn't say that we provoked the war and they had to retaliate.

The Japanese military had hundreds of spies in the Hawaiian Islands as well as in California. They sent reports back to Japan outlining all military activities so of course it was well known that the Navy was always on shore leave on the weekends and only a so-called "skeleton force" on duty on the ships and air bases.

Every Saturday night was party time and everyone generally slept late on Sunday mornings. Most had hangovers and weren't in full possession of their mental faculties.

When the Japanese planes were still several hours away from Pearl Harbor, a Staff Sergeant Radio Operator picked up the chatter and reported that a vast group of planes was headed for Hawaii. But his superior and the Navy each thought it was some activity planned by the other, that no one in their right mind would dare attack Hawaii!

And beside, there were Japanese envoys in Washington, DC, negotiating a trade agreement that the Japanese sorely needed. The standard story of the war was that we had shipped thousands of tons of scrap iron to Japan during the years prior to the Pearl Harbor attack. We sure got it back–in the form of bombers killing Americans in the Islands!

In my mind, I can still see pictures taken of the Emperor of Japan in full military uniform astride a beautiful white horse, which had been shown in various civilian publications. He was well aware of what was going on in his country and should have been tried with the others as a war criminal.

I'm sure that if the Japanese had wont he war, he would have ridden his horse in a victory march down Pennsylvania Avenue in DC, in front of all his troops.

If the Japanese military had known that most of the U.S. Navy lay at the bottom of Pearl Harbor and that there was mass confusion, they could have taken Hawaii easily. But their intelligence wasn't that good or fast enough.

I can see the Army and the Air Corps troops scampering out of their barracks and going to the ordnance building to check out rifles and ammo, because they weren't allowed to have weapons in the barracks. Oh yes, "get the serial number of each weapon and have the soldier sign for it!"

I felt sorry for the two lower ranking commanders charged with not having our military alerted before the attacks. But no one can do anything without orders from the high-ranking commanders in Washington, DC telling them to do so immediately. Those two men were "sacrificial goats".

I can remember when General Billy Mitchell was tried because he refused to shut up about the fact that the Army Air Corps should have long-range bombers. The Navy commanders all said that the Air Corps did not need them because the Navy was our first line of defense. Brigadier General Mitchell was court-martialed and reduced to the rank of Colonel and given a desk job as Commander of the Southwest Eight Army Air Corps detachments that consisted of small units in Texas, Arizona, New Mexico, and Oklahoma, which serviced military aircraft at civilian airports.

Gen. Mitchell's office was in the Quadrangle at Fort Sam Houston in San Antonio, Texas, in 1928. I was attached to that organization in 1937 and served in Tuscon, Arizona, El Paso and Brooks Field in San Antonio.

It must have been a comforting thought to him to recall that the Navy had anchored an old battleship that had been stripped of armament off the east coast in the Atlantic. It was to be used for target practice. General Mitchell sunk it with one 500-pound bomb down the smokestack! The Navy yelled "foul!" and claimed it was a "lucky hit"!

Later, Pearl Harbor proved Gen. Mitchell right. We never heard that motto, "The Navy is the first line of defense" again.

Gen. Mitchell was later restored to his former rank and pension.

In the early part of the Pacific War, when morale was low, Captain Colin P. Kelly was flying a bomber and was hit by Japanese naval fire. As he went down with his plane, loaded with bombs, he rammed it down the smokestack of a Japanese battleship and sunk the ship. That gave us old Army Air Corps people a huge lift!

Later, we flew to the beautiful lush island of Kauai, to a Sheraton Hotel located on twenty acres of ocean front gardens. Ruthie and I, along with two other people, boarded a jet helicopter for a ride over the area where the movie, "Raiders of the Lost Ark" was made. It was quite a jungle on each side of the stream. Then we flew over the area where "Jurassic Park" was filmed. The flight took us up the mountainside to the rim of a volcano, over the falls, and all around the beautiful area.

I asked the pilot what made the sheer cliffs so green. In answer, he flew the helicopter close to the cliffs, hovering several hundred feet above the stream below. It was almost as though I could reach out and touch the cliffs! He hovered so that we could see what made them so green.

It was covered with moss and ferns! They have over 500 inches of rain there a year—the wettest place on Earth. It was very beautiful and I enjoyed the flight so much that I wished it could last much longer. The helicopter was like a bubble and you could see down through the clear plates by my left foot to the ground. I loved it and wasn't afraid of the heights at all.

After a couple of days of tours and banquets, we were waiting at the hotel entrance for our bus to pick us up. An elderly gentleman sat down on the bench beside me and asked me how I liked his island. I replied that it was fantastic, but I wished I could have fished in the surf or one of the coves.

He laughed and said, "I've lived here all my 90 years and I would like nothing better than to take you to my cottage and show you some real fishing!"

I told him that we were leaving for another island as soon as the tour bus arrived. His daughter arrived to pick him up and I told her that I wished I could have fished with her father. "Dad is an avid fisherman," She said, "and he knows where to fish. He must really like you or he wouldn't have asked you to go fishing with him! He is a true Hawaiian and has many different nationalities in his family tree!

It was a sad farewell for me as I said goodbye to them.

We flew to Maui and had a free day to explore the sights on our own. We took a bus tour to the summit of Mount Haleakala, some 10,023 feet high. Inside the rim lay a sleeping volcano with a bowl seven miles across and one mile deep. The rocks and ask reminded me of the pictures of the moon. There were rocks both inside and outside the rim all the way down to the valley we had come from. We were instructed not to wander off the paved road and walkways because we could start a rockslide. And we were told how the islands were formed.

Since man inhabited the Hawaiian Islands, they have lost ten percent of the plant life, which is more than all we have in the United States of different species. It all began millions of years ago. The plant life drifted in on the sea or the wind or was brought by birds from other islands.

Indonesians, in their sixty-five foot long, twin hull boats, were the first to explore the Hawaiian Islands. Their canoes were fastened together side by side by wood beams that kept them about ten feet apart. These beams were covered with wood planking, forming a deck, and a thatched hut was built that covered most of the deck, forming a pontoon boat.

The front and rear sections of the canoes that weren't covered by the deck were covered with sail material to keep out the sea spray and waves. They carried their food with them–plants, pigs, nuts, fruit, and so forth. They also carried a good supply of fresh water for the voyage. The boat had only one mast and one sail, and a tiller to steer the boat.

In later year, pigs that had escaped had gone wild and destroyed plant life that is no longer found there or anyplace else in the world.

On our second day and evening, we joined a large group of people at an outdoor dinner theater, which was a true Hawaiian luau with roast pigs and all the trimmings and beautiful Hawaiian music playing in the background.

After the feast, we saw "The Drums of the Pacific" performed onstage with singers and dancers. The "Tai", a Hawaiian drink made with rum and pineapple juice, was smooth and very tasty, and helped us to get to know each other better. The lighting with torches and fires and colorful lights was unbelievably fantastic, more like a dream!

The next day we went shopping in town and looked at the gorgeous flowers, shrubs, and bynan trees. We dined that evening at the Hyatt's Swan Court, which had several waterfalls, ferns, and shrubs hanging down the cliffs. There were even several swans swimming in the large pool of water.

This was one of the ten most famous romantic places, according to the TV program, "Lifestyles of the Rich and

Famous". It was a fantastic place to spend the evening with our fellow travelers.

The following day, after a wonderful farewell, we flew home from Honolulu via Dallas and Atlanta, to Philadelphia. It was a great vacation!

In October 1997, friends of ours in Baltimore called to ask if we would be interested in a trip to Branson, Missouri. They were driving there for a music festival, which usually lasted all winter. Of course, we said yes, provided they let us share expenses. They agreed and off we went.

It was a wonderful trip to make in October, to see all the foliage changing color. Branson is a small town of only 10,000 residents, but it has 50 motels, 40 restaurants, and 30 deluxe theaters! Performers like Mel Tillis, Roy Clark, Andy Williams, Johnny Cash, and many more, own the theaters. These bands tour the country throughout the summer and spend the winters in Branson.

The town was swamped with buses from all over the country, just like the Atlantic City, New Jersey casinos are all year round. Besides the theaters, there were other attractions like the "Believe it Or Not" Museum, and the Stone Hill Winery, which is one of the oldest in mid-America. They bottled dozens of different types of wine from the vineyards in Arkansas and Missouri.

The most fantastic part of the tour of that area was our experience at a dinner theater aboard the Branson Belle, a paddlewheel boat. It lasted for over two hours and I didn't even realize we were cruising until I happened to look out through the portholes in the entrance doors and noticed the trees moving past! There was no engine sounds at all.

We had a nice table where we had a good view of the stage but weren't too close to the orchestra. The food was Cajun and so were the performers. We loved it and were very thankful that Louis and Florence Loewner had invited us along!

CHAPTER TWENTY-THREE

Medical and Physical Problems, But Still, a Great Retirement!

For months, I had been bothered by indigestion. I would take antacids and was fine again. I continued taking my walks and bowled as usual and the pain would vanish. My six-month checkups were always perfect. The blood work was excellent, my blood pressure perfect. Even my EKG's were fine. I continued to get the burning in my chest whenever I walked too fast, but after a short pause, I was always ready to walk again.

Ruthie finally took me to see my cardiologist, Dr. Goomburg. He gave me a stress test and found that I had three blocked arteries. He immediately sent me to a hospital in Center City, Philadelphia, and the following day, in February 1996, I was in surgery for a triple bypass.

Everything that could go wrong did, from a staph infection to pneumonia and I was placed on a ventilator. They had to reopen the incision to remove the staph infection and replace the skin and muscle grafts by plastic surgery to cover the former infected areas. It reached a point that my relatives were notified that I would probably not recover.

Had it not been for my faithful wife, I wouldn't have survived. Ruthie boarded a bus or senior citizen van every day to come see me. She knew that the ventilator was the only thing keeping me alive, and her daily visits proved to be a blessing.

One day, the ventilator's plastic nose piece had dug into the bridge of my nose and the tape holding it in place had slipped down, covering my nostrils. I had restraints on both arms, so I could not help myself, and was fighting for breath. Ruthie came into my room at the right time!

Being a former nurse and a medical assistant for twenty-three years, Ruthie immediately saw the problem. When she saw me struggling, to put it mildly, "all hell broke loose"! She "shook up" the whole place and my room was suddenly full of people trying to help!

From then on, I began to slowly recover and I was certainly given more attention both day and night. I felt no pain, except when the wounds were re-bandaged and cleaned. They still had my arms and legs in restraints so I wouldn't pull the tubes and IVs out.

I can remember thinking that the government was trying to kill me and make it look like an accident unless I talked. I must have had some sort of reaction to the pain medication! It was a continuous nightmare, in living color. I would wake and go back to sleep, falling back to the same dream. Perhaps I had seen too many movies and TV shows that dealt with those types of plots and intrigue.

During my military service and my civil service career, I had several security clearances, including "top secret", but at no time was I ever engaged in any undercover work.

I was in the hospital for forty days, and spent twenty-one of those days on the ventilator, and twenty-eight days in Intensive Care.

Upon my release, I was sent to the Chestnut Hill Rehab Hospital where I spent another twenty-eight days learning to walk, talk, and care for myself. They were a great bunch of well-trained technicians and had me out of the wheelchair in a few days and walking again, without a cane, in two weeks.

The exercise and therapy sessions were tough, but persistence paid off when I could walk again and take care of myself.

The only problem I had during that time was that my epiglottis (the piece of cartilage that closes off the windpipe when one swallows) wouldn't close and any liquid would go straight into my lungs. I had to be fed through a "belly tube". In rehab, they made it all portable so I could exercise and walk without dragging those bag holder stands on wheels with me.

Ruthie, my devoted wife, was there every day, making sure I had my voice, physical, mental, and occupational therapies, all of which began the day I arrived and continued even after I was sent home twenty-eight days later.

I had many visitors, which I appreciated, especially Ruthie's daughter, Jackie, her husband, Harry, and their two children, Laura and Ian. They all brought me cheer and a renewed vigor to keep trying to get well.

One incident in my room in rehab stays in my memory. Around 2:00 one morning, a nurse's assistant was going to put some medication on my "belly tube". She accidentally knocked over a glass of ice water sitting on the mobile tray beside my bed. The icy water landed on my "family jewels"! I was sound asleep and awoke abruptly, thinking that someone had set me on fire. I almost had a religious experience!

I yelled as loud as I could and within second, I had more help than I needed! That poor nurse's assistant was new and looked at me in horror, but later we became friends.

I was released with a large bag full of medication, bottle, etc., and became an outpatient at Chestnut Hill Hospital for the next six weeks. I also had a visiting voice therapist, Dr. Altschuler, who had me drinking water in a couple of weeks. She really knew how to do that, explaining to me to take a small sip and put my head forward a little before I swallowed. I knew I would be fine when I felt that ice water hit my stomach and not my lungs! A few weeks after that, after a lot of soft foods, like applesauce, yogurt, and Ensure, I had the belly tube removed.

I had a serious setback due to an enzyme that developed in my intestines, but a great doctor found a combination of medications that cleared that up.

In four or five months, I was back to bowling in two leagues again, using an eleven-pound ball, thanks to all the exercise in rehab that made me strong enough again. The people I bowled with were the kindest bunch of people. I had received a "hurry and get well" card with a hundred and twenty signatures on it! And the 15-pound ball I use now, at age 86, feels just fine. I'm bowling in three leagues a week now!

I'm back to driving in the city and on the highways, visiting Ruthie's and my friends often. And of course, the Diviak Family Reunion in Abingdon, Virginia is a must! It signifies a renewal that rekindles the warmth in my heart for my relatives, because we share something very special and stay in touch.

In February 1999, we lost both my sister Anna, who was 89, and my sister Helen, who was 87. Although they lived at least a thousand miles apart, they both died in their sleep on the same day!

My sisters Mary and Julia both died at age 75, a few years ago.

There are five of us left: I at 86, sister Teresa is 94, brother John is 82, and Steven is 80, and the "baby", Josephine, who is 77. It has been said that if you want to live into your eighties or better, picking a family tree whose ancestors lived to that age or older is a good start. However, because of the huge strides medicine has made, the odds of anyone living that long is greatly improved.

In June 1999, I had microsurgery on my left eye to remove a cataract and a month later, the right eye was done. Prescription lenses were implanted in both eyes. The left eye was quickly done, only about twenty minutes, but the right eye took a bit longer.

I'd suffered a trauma to that eye back in junior high football and had forgotten about it. After all, it was seventy years ago! Besides, I had a flying physical six years afterward in order to get into the Army Air Corps and passed with 20+15 vision in both eyes. This was a surprise to me. And there were many other physicals but the trauma to my right eye was never discovered.

When I was in seventh grade at Coalwood Junior High School, the assistant coach asked me to go out for football. I did and was made a halfback. The only thing that bothered me was that the largest helmet they had available was too tight for me, but by stretching the leather and elastic parts, I was able to wear it.

During the first play of the first game, the center snapped the ball to me (no "T" formation in those days!) and I took off at full speed. I thought I was in the clear when suddenly I felt a slap in my right eye. It stung like a needle prick and I went down with blood in my eye. A long fingernail of the safetyman had gone up under my eyelid and cut the iris.

I was rushed to the coal company's doctor there in Coalwood. He cleaned the eye, applied a thick compress, and let me go to catch the school bus back to Caretta. Once back in Caretta, the bus dropped me off at the company store and I walked the mile to my home. My parents were upset and vowed no more football for me.

The Coalwood doctor had called our company doctor in Caretta the next day to have him check on me and change the dressing. He also recommended the Caretta doctor speak to my father and tell him "Joe should not play football that year."

I followed the doctor's orders and slept propped up in bed for a few days. The pain was there all the time but after the eye opened and the swelling diminished, I felt better. But I had a bloodshot eye for a few weeks.

That ended my football career, much to my disappointment. But it amused me that I couldn't play football because it was "a dangerous game", but working in the Caretta Shaft Mine, which went 650 feet straight down and extended six miles under the mountain was fine!

My parents thought that all I was doing at the mine was construction work, which included loading slate and rock in the mine cars when I started the summer before I went into the tenth grade. I never told them that I was operating a jackhammer, drilling rock in the tunnels, ceilings, and slate bottoms, then doing the blasting because I was one of the few "lucky ones" who wasn't bothered by the dynamite fumes!

I never told them of all the narrow escapes either. All I wanted to do was work so I could get out of there permanently. I knew I needed that piece of paper that said I had graduated high school and all the knowledge that went with it.

These memories all came back to me with a rush when I was told of that old injury, or trauma as it was called, to my right eye.

Most of us coalminer's children learned to live with pain. There were very few painkillers available to us. Old home remedies were often used. It wasn't until the 1930's that we could get over the counter drugs in the nearby cities of War and Welch. The doctors did have laudanum, a mixture of alcohol and morphine, but they seldom used it. I liked my mixture the best—a large soupspoon of corn whiskey and sugar right before I went to bed. I was glad I had very little pain!

The company doctors had their hands full with mine-related injuries and made house calls for emergencies only.

Midwives delivered most of the babies born then. All nine of us Diviak children were delivered that way.

Dr. Andrew Phillips, my ophthalmologist, did a wonderful procedure on both eyes. I was astonished at how beautiful all the colors were; they were all so bright and not just shades blending into each other. There were no more halos around lights and the moon. And best of all, I no longer felt as though I was looking out a window—I no longer needed glasses. I could see everything now without them and what a beautiful panoramic view I could see by merely moving my eyes and not turning my whole head.

It is a beautiful world.

Ruthie and I have been around for more than eighty-six years and we attribute that to being very active every day. Ruthie is now and has always been a beautiful, trim, and charming lady. Everyone that meets her likes her instantly and cherishes her friendship. Our bowling buddies, men and women, let me know often that I'm the luckiest man in the world to have Ruthie as my wife. Of course, I agree with them. She is a very caring and loving wife.

We bowl in three leagues, walk a mile each day, and if the weather is not conducive to going outdoors, we use our treadmill and exercise cycle in our cozy basement room in front of the TV. I also use a spring-loaded gadget every day to do my exercises, which has enabled me to bowl better. And, oh yes, I feel better each day. It seems that all this is necessary to keep these "two thousand body parts" functioning normally and the "corps suckles" circulating so they don't settle to the bottom, sort of like salt in a saltshaker!

We both still drive, both in the city and on our trips, changing drivers every two hours, which works well for us.

In correspondence with my old Army and civil service friends, they have all remarked, "Why don't you write a book? It should be a good one!"

I'm sorry I didn't do this sooner. There are still a few of them around, enjoying their retirement. They were all wonderful friends and were always there when I needed them.

I know that the two hundred or more descendents of the Diviak family will have the genes and desire to achieve the best for their families, attributes passed down by their Slovakian ancestors.

All of the Diviak descendants I have met seem to be very ambitious and are not satisfied to merely maintain a meager existence. They have all strived to do better.

Hopefully, the family tree and the minutes of the Diviak Family Reunion held every two years would always be available to them. 2003 will be the 20th anniversary of the first Diviak Family Reunion. The first reunion was held in 1983 at the home of my brother John and his wife, Frances in Abingdon, Virginia.

My oldest daughter, Ellen, is a retired high school history teacher and still lives in Bryan, Texas. Her husband, Hubert, is still an engineer and land acquisition/restorer for a strip mining company in Texas and will retire in a few years. Their son, Brooks, is a graduate of Texas A&M, like his father, and is an instructor in ballistic missiles in California. Brooks served five years in the Air Force and is a captain in the Reserves. He married Kimberly Horton while he was stationed at Vandenberg AFB. They have two children, Jackson, who is five years old, and Presley, who is two.

Ellen's daughter, Wendy Lyn, is a graduate of North Texas University, where her grandmother, my first wife, LaRue, graduated in 1939. Wendy is a certified art teacher but is presently a "stay-at-home mom". She is married to Kelly, a Texas A&M graduate, who is an architectural engineer and heads a large firm in Dallas.

Wendy and Kelly have three children of their own, Victoria, who is six, and Luke, who is three, and Jake, who is six months old, and two children from Kelly's first marriage: Kyle is seventeen and Amanda is fourteen. They are all doing well in school.

My daughter Barbara is an elementary school teacher and like Ellen, graduated from St. Mary's University in San Antonio and has a Master's Degree. Barbara still has a few years to go before she retires.

Grandchildren with Joseph and Ruthie Diviak in center (1999)
Brooks E. Nelson holding Presley on the left, and Kimberly K. with Jackson E. Nelson in front of her.
Vandenburg Air Force Base, California (8/29/89)

My children in 1989 in Bryan, Texas
Seated: Wendy, my daughter, Ellen Nelson, her mother, and Barbara J. Diviak, my daughter
Standing: Hubert D. Nelson, and Brooks E. Nelson, his son

My grandchildren, Carrollton, Texas (1998)
Back row: Granddaughter Wendy N. Osburn, her husband Kelly, and his two children Armanda and Kyle Osburn.
Front row: Wendy's and Kelly's children, Luke and Victoria Osburn.

This is my Diviak family. I am very proud of their accomplishments and their resolve to always "do better". It is for them that I write this book, so that they and their children will have some of the history of their ancestors.

A few years ago, a friend asked me what I would like to have inscribed on my tombstone. The question took me by surprise but after a moment, I replied that I liked what Supreme Court Justice Thurgood Marshall's reply to a TV reporter who had asked him that very question: "He did his best."

To me, that statement fits my continuous battle for success over all the "chuckholes" of life that I experienced on my road to the fulfillment of my dreams and a wonderful retirement.

The obstacles I had to overcome and the narrow escapes I survived were an education in each encounter. I survived and it made me a better person. I became more tolerant and understanding of the many people with problems I encountered in my 86 years of life and over fifty times I moved from one place to another.

I helped every one of them all that I could and was rewarded many times over in each instance. That is the reason I could write this book with love and humor, with truthfulness and not rancor or bitterness or, putting it crudely, "to get even" with someone.

That wonderful admonition to "love your neighbor as you would thyself" has always been my motto and way of life; even when that neighbor wasn't too lovable, I tried\, and I made a lot of friends from it.

By loving yourself, you radiate a positive attitude that draws others to you and they like it instantly. But the secret is to not overdo it—loving oneself too much is one of the worst things that can happen to a person who wants to be successful in life.

When I had "a mountain to climb" along the way in my life, I can still hear the voice within me that I heard when I was just eight years old and had seen two men after they had been crushed to death in the coalmine. In response to my fervent prayers in my secret spot on the mountainside, the voice said: "Joe, do your best and you will be all right! "

It is still a comfort to me because I know I did my best.

In the conclusion of this book, I have a parable of my own:

As we reach the twilight years of our lives, always remember that in the cup of life, the sugar is always in the bottom of the cup. The sweetness depends on the loving kindness we shared with others along the way getting there.

"The Hudson"

The ship that Joseph E. and Helen Kalny Diviak escaped from Europe on to America, October 20, 1906.

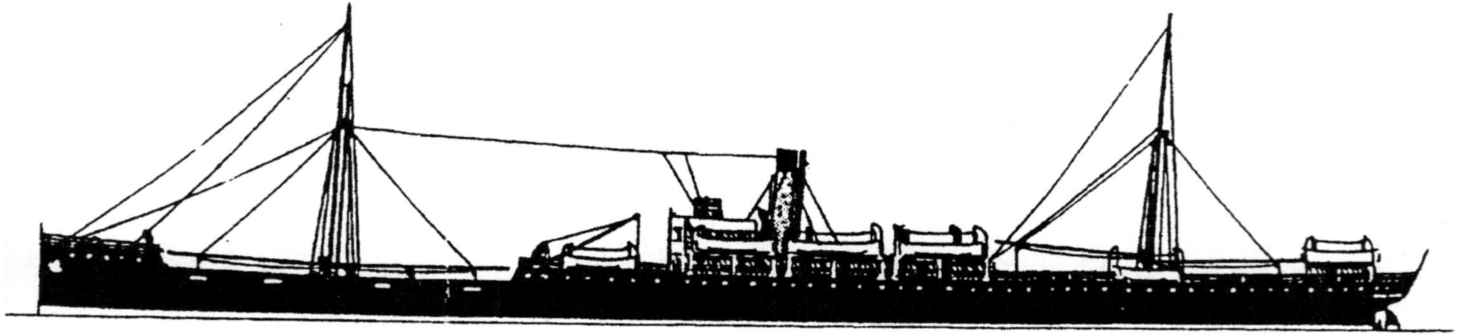

Built by Chantiers de Normandie, Grand Quevilly, France, 1905. 5,558 gross tons; 390 (bp) feet long; 50 feet wide. Steam triple expansion engine, single screw. Service speed 12 knots. 760 passengers (60 second class, 700 third class).
Built for French Line. French flag, in 1905 and named **Hudson**. Le Havre-New York and later Le Harve-Montreal service. Scrapped in Belgium in 1930.